Problem-Solving Experiences in Mathematics

ANNE M. BLOOMER

RANDALL I. CHARLES

FRANK K. LESTER, JR.

TEACHER SOURCEBOOK

GRADE K

DALE SEYMOUR PUBLICATIONS

Pearson Learning Group

Managing Editor: Cathy Anderson
Project Editor: Mali Apple
Production: Leanne Collins
Design Manager: Jeff Kelly
Text and Cover Design: Christy Butterfield
Illustrations: Joan Holub

Printed in the United States of America
ISBN 0-201-49087-0
8 9 10 11 12 06 05 04 03

CONTENTS

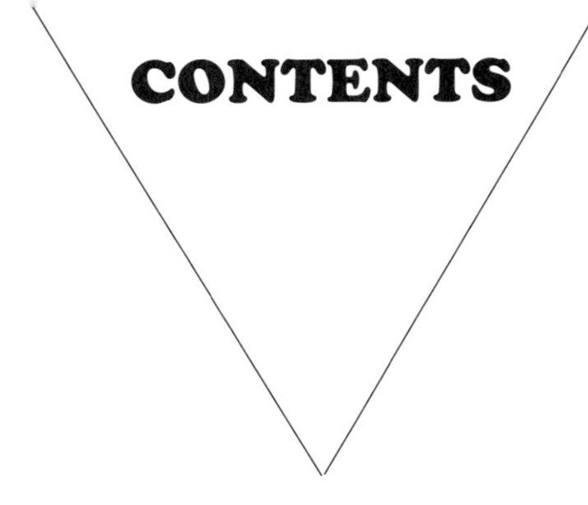

Overview

Problem Sets

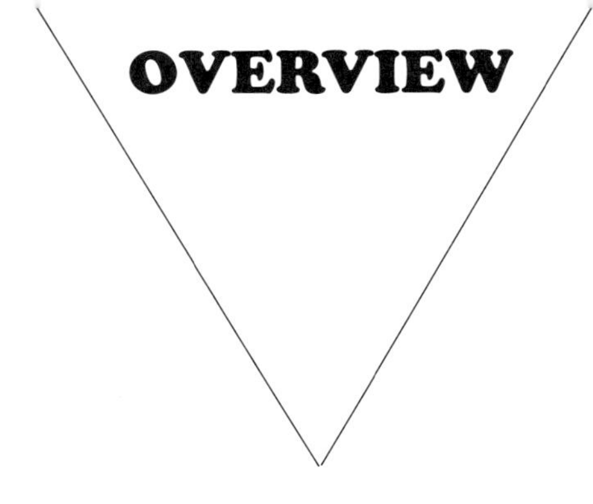

"The importance of problem solving to all education cannot be overestimated. To serve this goal effectively, the mathematics curriculum must provide many opportunities for all students to meet the problems that interest and challenge them and that, with appropriate effort, they can solve."

—*NCTM Standards, 1989*

P*roblem-Solving Experiences in Mathematics* (PSEM) provides the kinds of opportunities described in the NCTM Standards and is designed to supplement any instructional program. It consists of 60 problem-solving experiences grouped by story themes and a teaching strategy for problem solving. A separate package of blackline masters provides students with support material.

PSEM was designed with the NCTM Standards in mind.

Goals of the Program

The ultimate goal of any problem-solving program is to improve students' performance at solving problems correctly. Although this is the ultimate goal, instructional goals need to be more specific and developmental. The goals of PSEM are to

1. Improve students' willingness to try problems and improve their perseverance when solving problems.
2. Improve students' self-concepts with respect to their abilities to solve problems.
3. Make students aware of problem-solving strategies.
4. Make students aware of the value of approaching problems in a systematic manner.
5. Make students aware that many problems can be solved in more than one way, including with the use of manipulatives.
6. Improve students' abilities to select appropriate solution strategies.
7. Improve students' abilities to implement solution strategies accurately.
8. Improve students' abilities to monitor and evaluate their thinking while solving problems.
9. Improve students' abilities to get more correct answers to problems.
10. Improve students' abilities to communicate their thinking.

Organization of the Program

There are three types of problem-solving experiences in this program: problem-solving readiness activities, problem-solving skill activities, and process problems. The 60 experiences in this book are grouped into 15 sets. Each set begins with an introductory story that provides a unifying theme for the experiences in that set.

Problem sets are built around real-world themes.

PSEM provides a problem-solving experience—60 in all—twice a week during the school year. The problems were selected and sequenced so the concepts and skills needed to solve each problem would have been introduced to students approximately two months before they are encountered here, *if* the teacher follows the scope and sequence of lessons in most textbooks. For problems at the beginning of the year, concepts and skills are limited to those most students should have encountered prior to kindergarten. This organization means that students' work is limited to a *review* of concepts and skills. The emphasis on problem-solving instruction can thus be on understanding problems, selecting and implementing appropriate solution strategies, and checking one's work, rather than on carrying out computational skills.

PROBLEM-SOLVING READINESS ACTIVITIES are experiences designed to prepare students for future problem-solving experiences by building their confidence in dealing with real-world situations that involve numbers. Six types of readiness activities are included in Grade K:

1. Tell how numbers are used in the real world.
2. Given a story, answer questions about information in the story.
3. Tell a story using a given number.
4. Given a story, visualize objects and action in the story.
5. Given a number story, retell the story changing the numbers or the setting.
6. Given a story, act out the action in the story.

PROBLEM-SOLVING SKILL ACTIVITIES are experiences designed to promote the development of thinking processes involved in problem solving. Five types of skill activities are included in Grade K:

1. Given a number story, tell a question that can be answered using data in the story.
2. Given a picture, choose/tell a questions whose answer would be found by using addition/subtraction.
3. Given a story problem and the start of a picture, complete the picture to match the story problem.
4. Given the story problem with missing data, choose/tell appropriate data for solving the problem.
5. Given a story problem, tell whether addition or subtraction is needed to find the solution.

Problem-solving readiness and skill activities build a foundation for problem solving.

PROCESS PROBLEMS are problems that cannot be solved by simply choosing an operation. Instead, process problems are solved using one or more of these strategies:

1. Guess and check
2. Draw a picture
3. Make an organized list
4. Make a chart or graph
5. Look for a pattern
6. Use logical reasoning
7. Use manipulatives

Because process problems cannot be solved by simply choosing an operation, they exemplify and provide practice with the thinking processes inherent in problem solving. The following chart indicates the strategies that can be used to solve the process problems in this book. The chart does not show all of the possible ways of solving the problems—only those that are most commonly used. Also, manipulatives can be used with any of the strategies; their use is incorporated within all of the strategies.

Process problems exemplify and provide practice with the thinking processes inherent in problem solving.

LIKELY STRATEGIES USED IN THE PROCESS PROBLEMS

GUESS AND CHECK

3–4, 15–16, 39–40

DRAW/USE A PICTURE

27–28, 51–52

MAKE AN ORGANIZED LIST

35–36

MAKE A TABLE OR GRAPH

31–32, 55–56

LOOK FOR A PATTERN

7–8, 19–20, 43–44

USE LOGICAL REASONING

11–12, 23–24, 47–48

The two process problems in each set are matched by probable solution strategies to enable you to teach students how to use problem-solving strategies. Even though the process problems are matched within a set, *it is very important that students are not forced to use the strategy suggested by the hint. In fact, students should be encouraged to find solutions to problems using as many different strategies as they can.*

The process problems in Grade K are organized in the following manner.

ORGANIZATION OF THE PROCESS PROBLEMS

SET 1
GUESS AND CHECK

SET 2
LOOK FOR A PATTERN

SET 3
USE LOGICAL REASONING

SET 4
GUESS AND CHECK

SET 5
LOOK FOR A PATTERN

SET 6
USE LOGICAL REASONING

SET 7
DRAW/USE A PICTURE

SET 8
MAKE A GRAPH

SET 9
MAKE AN ORGANIZED LIST

SET 10
GUESS AND CHECK

SET 11
LOOK FOR A PATTERN

SET 12
USE LOGICAL REASONING

SET 13
DRAW/USE A PICTURE

SET 14
MAKE A GRAPH

SET 15
MAKE AN ORGANIZED LIST

Building a Positive Classroom Climate

In the first two months of the school year, the most important goal with regard to problem solving should be to establish a positive classroom climate. Then, you can begin to focus on the development of the students' problem-solving abilities. The importance of a positive classroom climate cannot be overemphasized in building a successful problem-solving program.

In the first two months of the school year, establish a positive classroom climate.

Many factors affect classroom climate. Among the most important are the appropriateness of the content (not too difficult and not too easy), the teacher's evaluation practices, and the teacher's attitude and actions related to problem solving. Of these, the teacher's attitude and actions are most important. Here are some things that will help you establish a positive climate in your classroom for problem solving:

- Be enthusiastic about problem solving.
- Have students bring in problems from their personal experiences.
- Personalize problems whenever possible (e.g., use students' names).
- Recognize and reinforce willingness and perseverance.
- Reward risk takers.
- Encourage students to play hunches.
- Accept unusual solutions.
- Praise students for getting correct solutions, but during problem solving, emphasize the selection and use of problem-solving strategies.
- Emphasize persistence rather than speed.

The Teacher's Role

All of the problem-solving experiences in this book were designed to be given orally, with the teacher playing an active role leading the students through each experience.

USING THE BLACKLINE MASTERS Many of the problem-solving experiences in this book are accompanied by a blackline master (BLM). BLMs provide support material for the problem-solving experiences; *they are not worksheets to be assigned to students.* BLMs can be used only as part of the teacher-oriented, oral problem-solving experiences. A sample BLM for a process problem is shown here. The students should have a copy of each BLM.

USING THE INTRODUCTORY STORIES These stories provide a unifying theme for all of the problem-solving experiences in a set. Each story has an accompanying BLM to be distributed to the students. The stories are to be read by the teacher as the students look at the BLM. The discussion questions given in the teacher notes can be interspersed throughout the reading of the story or can all be asked after the story has been read. The questions do not require mathematics to be answered. Instead, they help familiarize students with the theme, promote the improvement of students' listening skills, and promote the development of students' creativity. The introductory story and discussion questions also serve to motivate students for the problem-solving experiences in the set that follows. If you use a "problem-of-the-day" approach for this program, the introductory story could be read and discussed on Monday and the follow-up problem-solving experiences used and revisited throughout the next two weeks.

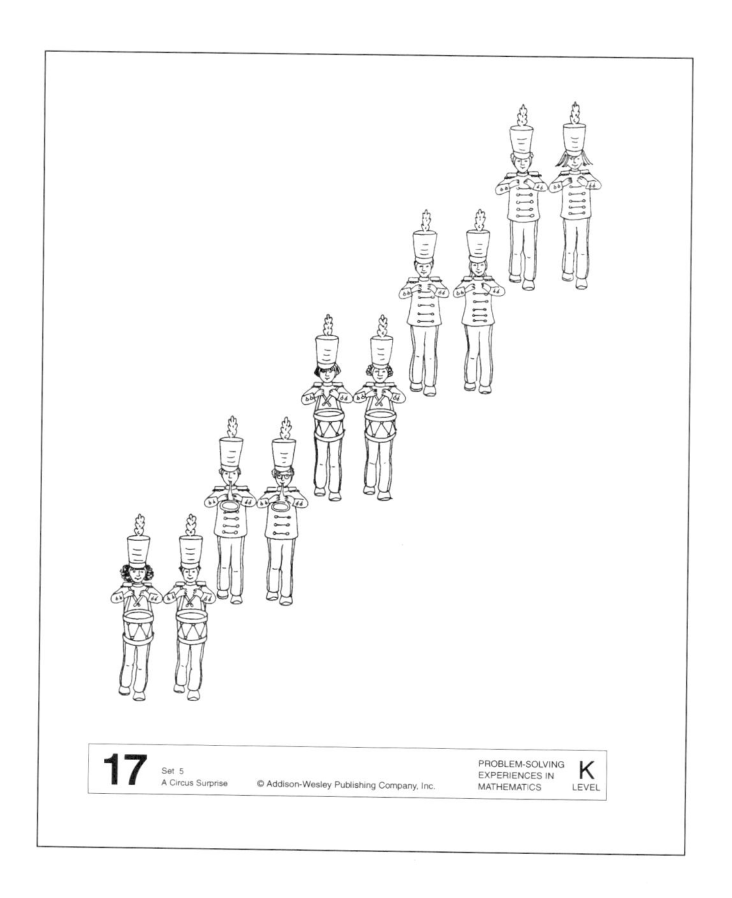

USING THE LESSON PLAN FOR PROBLEM-SOLVING READINESS AND SKILL ACTIVITIES The lesson plan for readiness and skill activities consists of recommended teaching actions specific to each activity. A sample activity is shown here. The teaching actions provide guidelines for how to (orally) introduce the activity and lead students through a discussion of the activity.

25 Readiness Activity

Retell a Number Story

STORY

Mai was passing out treats to the class. She passed out 4 heart-shaped cookies and 5 that were shaped like stars.

TASK 1

Retell the story with a different number of cookies.

TASK 2

Retell the story with stickers instead of cookies.

TEACHING ACTIONS

1. Read and discuss the story.
2. Give Task 1 to the children. Model it, then solicit stories with different numbers of cookies in them.
3. Repeat for Task 2.

USING THE LESSON PLAN FOR PROCESS PROBLEMS The lesson plan for each process problem is outlined as in the sample shown. Next to each section of the lesson plan are general teaching actions recommended for problem solving. The table on page xi gives a complete description of the teaching actions and describes the purpose of each.

TEACHING ACTIONS BEFORE

1. Read the problem.
2. Ask questions for understanding the problem.
3. Discuss possible solution strategies.

TEACHING ACTIONS DURING

4. Observe students.
5. Give hints as needed for solving the problem.
6. Require students to check back and answer the problem.
7. Give problem extension(s) as needed.

TEACHING ACTIONS AFTER

8. Discuss solution(s).
9. Discuss related problem(s) and extension(s).
10. Discuss special features as needed.

19 PROCESS PROBLEM

After the acrobat act, a very small car came out and drove to the center of the ring. The driver got out, and guess who it was—the clown who had surprised Josette! He was holding a silver and green balloon just like hers. Then clown after clown climbed out of the tiny car. Soon there were 6 clowns standing in the ring. Josette couldn't see what the last 2 looked like. Josette noticed her clown had a red hat and the next clown, with the dog, had a yellow hat. The third clown, with a feather, had a red hat. The fourth clown, with a purse, had a yellow hat. (Have children place counters on the appropriate hats.) Marie said, "Look at the colors of the hats. They make a pattern." What color hats were the last 2 clowns wearing?

MATERIALS

red and yellow 1" counters, tiles, or 1" construction-paper squares (4 of each per child or pair); red and yellow crayons

Understanding the Problem

- What were the clowns doing? (*getting out of a car*)
- Which clowns couldn't Josette see? (*the last 2*)
- Whose hat colors could she see? (*the first 4 clowns*)
- Why do we know there's a pattern? (*Marie said the hat colors made a pattern*)
- What do we need to find? (*the color of the hats of the last 2 clowns*)

Solving the Problem

- What color is this hat? (point to the clown with the balloon) (*red*)
- What color is the next hat in line? (point to the clown with the dog) (*yellow*)
- What color is this hat? (point to the clown with a feather) (*red*)
- What color is the next hat in line? (point to the fourth clown, with the purse) (*yellow*)
- Can you figure out the last two hats? Put counters on the hats and check to see if you have a color pattern for all the clowns' hats. Then color the hats the color of the counters.

16 Set 3 A Circus Surprise © Addison-Wesley Publishing Company, Inc. PROBLEM-SOLVING EXPERIENCES IN MATHEMATICS K LEVEL

Solution

Look for a Pattern

Pattern: (hat colors from left to right) red, yellow, red, yellow, red, yellow

Clown with balloon: red; clown with dog: yellow; clown with feather: red; clown with purse: yellow; firefighter: red; clown with trumpet: yellow.

Related Problems: 8, 7

STRATEGY ASSESSMENT IDEAS

Listen and watch as children work to see if they

- describe a pattern formed by the colors of the hats
- extend the pattern correctly with yellow and red counters
- use the pattern to arrive at a correct solution to the problem

Problem Extension

Have children look for other patterns in the line of clowns: buttons/no buttons; small shoes/large shoes.

28

We have found this strategy (i.e., the ten teaching actions) to be a valuable and easily learned plan for facilitating students' thinking and problem-solving work. A scenario is useful to illustrate how to use the teaching actions.

Before students start work on a problem, have a whole-class discussion about the problem, following Teaching Actions 1, 2, and 3. After this discussion, have students begin working on the problem. **During** the time they are working on the problem, move around the room monitoring and directing students' work (Teaching Actions 4, 5, 6, and 7). **After** they have solved the problem, have another whole-class discussion about the students' work (Teaching Actions 8, 9, and 10).

One of the key elements in successfully guiding students' problem-solving experiences is asking the right questions at the right time. For each problem, questions and hints are given in the lesson plan. The first set of questions (*Understanding the Problem*) should be used **before** students start work when you are helping them understand the problem (Teaching Action 2). The second set of questions (*Solving the Problem*) should be used **during** the time students are working on a problem, if or when they get stumped in their solution attempt (Teaching Action 5). The hints given for Solving the Problem should be viewed as *possible* hints.

Teaching Actions	Purpose of Teaching Action
BEFORE	
1. Read the problem to the class. Discuss words or phrases students may not understand.	To illustrate the importance of reading problems carefully and to focus on words that have special interpretations in mathematics.
2. Use a whole-class discussion about understanding the problem. Ask questions to help students understand the problem. (See the problem-specific hints for *Understanding the Problem.*)	To focus attention on important data in the problem and to clarify parts of the problem.
3. Ask students which strategies might be helpful for finding a solution. Do not evaluate students' suggestions. You can discuss the list of strategies on the problem-solving guide when asking for suggestions. (See page xiv.)	To elicit ideas for *possible* ways to solve a problem.
DURING	
4. Observe and question students about their work.	To diagnose students' strengths and weaknesses related to problem solving.
5. Give hints for solving the problem as needed. (See the problem-specific hints for *Solving the Problem.*)	To help students get past blocks in solving a problem.
6. Require students who obtain a solution to check their work and answer the problem.	To require students to look over their work.
7. Give a problem extension to students who complete the original problem much sooner than others. (See the *Problem Extension* section.)	To keep all students involved in a meaningful problem-solving experience until others have completed work on the original problem. (This is a classroom management teaching action. See Teaching Action 9 for using problem extensions to improve problem-solving ability.)
AFTER	
8. Show and discuss students' solutions to the original problem. Have students name the strategies used. You can reinforce the names of the strategies by pointing out the strategy names on the problem-solving guide. (See page xiv.)	To show and name strategies for solving the problem.
9. Relate the problem to previous problems (if possible) and solve an extension of the original problem. (See the *Related Problems* and *Problem Extension* sections.)	To demonstrate that problem-solving strategies are not problem-specific and to help students recognize different kinds of situations in which particular strategies may be useful.
10. Discuss special features of the original problem, if any. (See *Notes.*)	To show how special features of problems (for example, picture accompanying the problem statement) may influence students' thinking.

As you observe and question students, you must decide which, if any, of those hints are appropriate. Sometimes none of the hints listed will seem appropriate, and you will need to come up with others. Quite often you'll find it necessary to repeat one or more of the *Understanding the Problem* questions you used in the whole-class discussion **before** students started work. Most teachers find that selecting just the right hint for a student or group is a teaching skill that develops with experience.

In the *Solution* section, at least one solution to the problem is shown. The names of the solution strategies are given, and the answer to the problem is given in a complete sentence. The solutions shown for each problem were selected because they are ones used most often by students in our work with these problems. However, it is possible that students will use solution strategies different from the ones shown. That's fine! *Students should not be required to use a particular solution strategy for a given problem. Rather, they should be encouraged to find as many ways as possible to solve problems.*

The *Related Problems* section identifies (by number) problems that appeared earlier in the book that can be solved in ways similar to the given problem. The *Problem Extension* section includes an additional problem that is similar to the original problem. The answer to the problem extension is provided after the problem statement. Some problems have a *Comment* section containing an observation about the problems that could be used with Teaching Action 10 (discuss any special features of the problem). For example, some pictures accompanying problem statements can be misleading, and a statement to this effect could appear in this section.

TIME NEEDED TO IMPLEMENT THE PROGRAM This chart shows the amount of time you might expect to spend on the introductory stories and on each type of problem-solving experience if you use the complete set of teaching notes. It is important to realize that the time needed for each experience will be greatest at the beginning of the year.

Type of experience	*Approximate time required*
introductory story	10 to 15 minutes
readiness and skill activity	5 to 10 minutes
process problem	15 to 20 minutes

One of the key elements in successfully guiding students' problem-solving experiences is asking the right questions at the right time.

Cooperative Learning for Problem Solving

The introductory stories should be handled as a whole-class discussion. We recommend that students also work as a class on readiness activities and skill activities. For process problems, we recommend small-group work. Since process problems are usually challenging, small-group work helps reduce the pressure on the individual student, and it provides a structure for more easily monitoring and assessing the problem-solving performance of all the students in the class. Third, small-group work often elicits behaviors, such as justifying and evaluating ideas, that promote the improvement of problem-solving performance.

One of the important aspects in building successful cooperative group work for problem solving is to establish a class set of guidelines for working in groups. Consider devoting one class session at the start of the school year to developing a class set of guidelines. Invite students, working in groups, to discuss what they believe should be appropriate guidelines. Then, as a whole class, agree on a final list. Make a bulletin board displaying the guidelines, and refer to them on a regular basis before and after group work. Try not to have too many rules. Here are some guidelines students have suggested.

POSSIBLE SMALL-GROUP RULES

- Include everybody in the group.
- Share ideas.
- Talk only to your group.
- Participate.
- Cooperate with others.
- Pay attention.
- Be polite and kind to others.
- Listen.
- Follow directions.
- Talk quietly.
- Disagree when appropriate.
- Ask group questions only.

Using Manipulatives

Building and using models for important mathematical ideas helps most people to understand mathematics better. Models in the early grades are primarily *manipulative materials.* Manipulatives play a key role in this program at the kindergarten level and in many experiences at grades 1 and 2.

Suggested materials and tips for their use are provided in the lesson plans. The following manipulatives and other materials are called for in this book:

1" colored counters
1" tiles
pattern blocks
interlocking cubes
plastic links
construction-paper squares (*optional*)
paper clips
lengths of yarn (*optional*)
crayons

In many cases, alternative manipulatives can be used. For example, round 1" counters are commercially available, but other materials—such as 1" construction-paper squares, square tiles, buttons, or any other small counting objects—can be substituted for counters, depending on the activity. Alternatives for square 1" color tiles include 1" construction-paper squares and pattern block squares. Commercial plastic links are $1\frac{3}{4}$" long. If length is not a consideration for the activity, other materials that link or snap together can be used instead. The key to selecting a manipulative is its intended function in the problem.

Using manipulatives to solve problems may need to be modeled by you or other students to help all understand how the objects can be used to model mathematical situations.

Assessing Students

We have found three types of assessment tools particularly helpful in an assessment plan for problem solving: observations of student work, analyses of written work, and problem-solving portfolios.

> *An assessment plan for this program should not be limited to a check for correct answers.*

OBSERVING AND LISTENING TO STUDENTS Every assessment plan for problem solving must be built on the observations you make as you watch and listen to students as they work. PSEM provides three kinds of support in this area. First, every process problem provides *Strategy Assessment Ideas,* actions and statements related to the implementation of problem-solving strategies to watch and listen for as students solve problems. Second, a *Strategy Implementation Checklist* is provided in the Assessment Appendix (see page 95). This list can be used to record students' progress over time in their ability to appropriately use strategies. Third, a general *Problem-Solving Observation Checklist,* also in the Assessment Appendix, includes general problem-solving behaviors and dispositions to be observed and analyzed over time.

ANALYZING WRITTEN WORK An alternative to observing and listening to students as they solve problems is to use a holistic system for assessing written work, including students' written solutions to problems. The Assessment Appendix includes a five-level *Focused Holistic Assessment Rubric.*

USING PORTFOLIOS A portfolio is a collection of student work. Portfolios can be used for many purposes, with the intended purpose determining the contents. A common use of portfolios is to provide a collection of student work that can be analyzed for growth over time. The work may be chosen by the student, by the teacher, or by both. The Assessment Appendix includes a *Mathematics Portfolio Profile Checklist* for analyzing a student's portfolio for growth over time.

Here are some important things to keep in mind as you build your assessment plan for problem solving.

1. Assessment is not synonymous with grading.
2. An assessment plan should provide data useful in making instructional decisions.
3. All assessment plans should include observations and questioning of students.
4. Assessment should not be based on a single experience, but should look at student growth over time in a variety of kinds of experiences.
5. Every student does not have to be assessed in every problem-solving experience.
6. Assess thinking processes as well as the correct answer.
7. Assess attitudes and beliefs as well as performance.
8. Inform your students of your assessment plan.

Some Special Considerations

VALUE OF THE INTRODUCTORY STORIES Since students are not asked to solve mathematics problems in the discussions of the introductory stories, it is possible to underestimate the value of these stories. At the early grades, the context of the story is a problem characteristic that significantly affects students' abilities to understand problems. Although we have found the themes in this book to be interesting and familiar to most children, omitting an introductory story could influence their opportunity for success on the problem-solving experiences in that set.

REORGANIZING THE PROBLEM-SOLVING EXPERIENCES Should you decide to use only some of the problem-solving experiences in this book, the organization of the experiences has implications for how you should select and use them. *Do not start in the middle of the book.* The problem-solving experiences have been sequenced from easy to difficult. Even if you do not start the program near the beginning of the year, you should still begin with the first problem-solving experience in the book and move sequentially through the program.

A PROBLEM-SOLVING GUIDE Many teachers find the problem-solving guide shown below to be a helpful instructional aid in implementing the teaching actions for one-step and process problems. Most teachers make a bulletin board out of the guide. The guide is particularly helpful for implementing Teaching Actions 3 and 8. For Teaching Actions 3 and 8, students use the guide when suggesting strategies that might be helpful in solving a given problem (Teaching Action 3) and in naming strategies actually used to solve problems (Teaching Action 8).

PROBLEM-SOLVING STRATEGIES GUIDE

- Guess and Check
- Draw a Picture
- Make an Organized List
- Make a Table
- Look for a Pattern
- Use Logical Reasoning
- Use Objects
- Use Addition
- Use Subtraction

DEVELOP AN ASSESSMENT PLAN THAT IS COMPLETE YET EASY TO USE No one assessment technique can capture all aspects of students' thinking related to problem solving. So, build an assessment plan that looks at student work in several ways (e.g., observations, portfolios, written work), but be careful that your plan is reasonable to implement in the time available. It is better to use two or three assessment techniques well than several techniques poorly.

A Visit to the Apple Orchard

The children in Mr. Rank's kindergarten class had been learning about apples. Now they were visiting an apple orchard. At the orchard, the children could see the rows and rows of trees with apples hanging from the branches. Liam said, "Look at the different colors of apples!" Some were red, some were green, and some were yellow. Ms. Macintyre, who worked for the orchard, explained that each kind of apple has its own color when it is ripe.

The men and women pickers had to stand on ladders to pick apples from the high branches. They put the apples into bags hung over their shoulders. Then they emptied the bags into baskets on the ground. Suzanne cried "Here comes the apple truck!" and the children watched as the pickers put their baskets into the back of the truck.

Next the class went to the building where the apples were sold. There were pieces of different kinds of apples on plates and each child could pick a piece to eat. Melissa chose a piece with yellow skin and said it was soft and sweet. Jacques wanted a crunchy one and Ms. Macintyre showed him a piece with bright red skin. It was so juicy, the juice ran down Jacque's chin! As the children got onto the bus, they agreed that it had been a fun and delicious trip.

Discussion Questions

1. What did the children in Mr. Rank's class visit? (*an apple orchard*)
2. What color apples did the children see? (*red, green, and yellow*)
3. How did the pickers reach the apples on the high branches? (*they stood on ladders*)
4. What kind of apple would you have chosen to eat?

1 READINESS ACTIVITY

Tell How Numbers Are Used in the Real World

Story

Sometimes people come to the orchard to pick their own apples and put them in bags. The bags are weighed to see how much the apples will cost. Some people buy many bags of apples so they can make applesauce or lots of apple pies.

Questions

1. How many cars can you see? (2) Can you find the license plates? Why do cars have numbers on their license plates? (*to identify them and their owners*)
2. Can you find the mailbox? The number on the mailbox is one hundred twenty four. Do you have a number on your mailbox or the building you live in? Why do mailboxes and buildings have numbers? (*so people can be found for visiting, deliveries, mail, etc.*)
3. Can you find the scale for weighing apples? What do the numbers on the scale mean? (*they tell how heavy the bags are*)
4. Can you find the sign with an apple on it? The sign tells us when the orchard is open. What do the 8 and the 5 tell you? (*what time the orchard opens and closes*)
5. Can you see any other numbers? (*a truck with a number 3 on it, and a price sign that reads 50¢*) What do you think they mean? (*the 3 tells which truck it is and the 50¢ tells how much each pound of apples costs*)

TEACHING ACTIONS

1. Read and discuss the story.
2. Using the sample questions, discuss how numbers are used in the picture.
3. (*optional*) Extend discussion to children's own experiences with scales, license plates, and other personal numbers.

2 READINESS ACTIVITY

Tell a Number Story

Story

Mr. Olafson is going to bake apple pies. He will need to pick 5 bags of apples. He wants to pick some extra just for eating.

Task 1

Tell a story about the apple orchard that has the number 2 in it.

Task 2

Tell a story about the apple orchard that has the number 3 in it.

TEACHING ACTIONS

1. Read and discuss the story.
2. Give Task 1 to the children and model it. Solicit a variety of stories. Encourage children to make up challenging stories.
3. Repeat for Task 2.

3 PROCESS PROBLEM

Yawanda enjoyed the orchard so much she asked her parents to take her back. Together they picked 3 bags of apples. Yawanda has decided she wants to work picking apples in an orchard when she gets bigger. When she got home, she drew a picture of an apple tree with many red apples on it. (Have children place a red counter on each apple.) Next she drew a bag. How many of her apples do you think it will take to fill the bag?

MATERIALS

red 1" counters or 1" construction-paper squares (12 per child or pair)

Understanding the Problem

- What did Yawanda put on the tree? (*apples*)
- What are we using for apples? (*red counters*)
- What does it mean to fill the bag? (*apples/counters should touch and come up to the top*)
- What are we trying to find? (*how many apples/counters will fill the bag*)

Solving the Problem

- How many apples are on the tree? (*12*)
- How many apples do you think will fit in the bag? Elicit guesses.
- How can you tell if your guess is right? (*move that many counters to the bag*) Have children move counters to the bag.
- If your guess wasn't right, can you make a new guess? How can you check? (*add more counters or take some away*)
- How many apples does it take to fill the bag? (*10*)

Solution

Guess and Check

It will take 10 apples to fill the bag. There will be 2 left over.

Problem Extensions

1. Have real bags, apples, and a food scale. Have children guess the number of apples that will fit in the bag and the weight of the full bag. Write estimates on the board. Have children check by filling and weighing the bag. Check the answer against the estimates.
2. Substitute pattern blocks for counters. Have children fill their bag with one kind of pattern block. Why are there different numbers of apples in the bags?

STRATEGY ASSESSMENT IDEAS

Listen and watch as children work to see if they

- make a reasonable first guess (a range of 6–12 apples is reasonable)
- make a second guess using what they learned from checking with the counters
- use counters to check their guesses

4 PROCESS PROBLEM

Melissa was playing at Yawanda's house. They decided to draw more apple bags. They each had 10 counters to use for apples. Yawanda chose red. Melissa chose yellow because she liked the yellow apple she ate at the orchard. (Have children count out 10 red and 10 yellow counters and place them on Blackline Master 4. The shaded circles are red.) Yawanda was finished with her bag first, so she put a 1 on it. Melissa put a 2 on hers. How many apples do you think it will take to fill Yawanda's bag? How many apples do you think it will take to fill Melissa's bag?

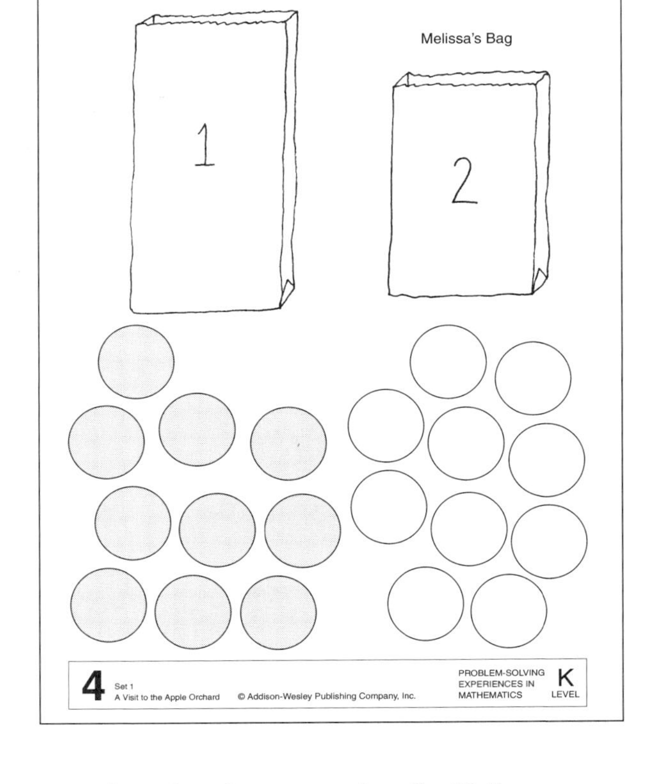

MATERIALS

red and yellow 1″ counters or 1″ construction-paper squares (10 of each per child or pair); red and yellow crayons

Understanding the Problem

- What are Yawanda and Melissa using the counters for? (*red and yellow apples*)
- What color did Yawanda choose? (*red*)
- What number bag is Yawanda's? (*1*) Have children outline bag 1 in red and color the apples under it.
- What color will go in bag 2? (*yellow*) Have children outline bag 2 in yellow and color the apples.
- What are we trying to find? (*how many it will take to fill each bag*)

Solving the Problem

- How many red apples do you think will fit in Yawanda's bag? Elicit guesses.
- How can you tell if your guess is right? (*move that many counters to the bag*)
- Was your guess right? If your guess wasn't right, can you make a new guess? How can you check your guess? (*add more counters, or take some away*) Have children add or remove counters.
- How many apples does it take to fill bag 1? (*8*)
- What else do we need to find? (*how many it takes to fill bag 2*)
- Do you think it will take the same number for bag 2? (*no, it is smaller*) Elicit guesses.
- How can you tell if your guess is right? (*move that many counters to the bag*)
- If your guess wasn't right, can you make a new guess? How can you check? (*add or take away counters*)
- How many apples does it take to fill bag 2? (*6*)

▶ *turn the page*

STRATEGY ASSESSMENT IDEAS

Watch and listen as children work to see if they

- make reasonable first guesses (a range of 5–12 apples is reasonable for bag 1; 4–7 for bag 2)
- make subsequent guesses using what they learned from checking with the counters
- use counters to check their guesses

Solution

Guess and Check

It takes 8 apples to fill Yawanda's bag (bag 1) and 6 to fill Melissa's bag (bag 2).

Related Problems: 3

Problem Extension

Suppose one of the girls used larger apples. Could she get more or fewer apples in her bag? (*fewer*) Have children check using larger manipulatives.

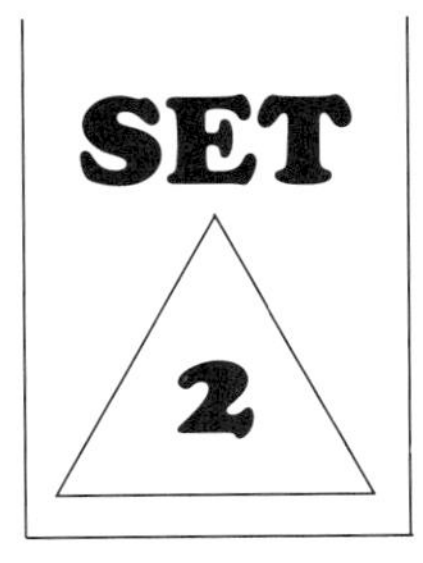

Favorite Stickers

All the boys and girls in Ms. Tsai's kindergarten class loved stickers. Melody collected dinosaur stickers. She even had three different sizes of Tyrannosaurus Rex! Monte liked the scratch-and-sniff kind. When he brought his sticker book in for sharing time, all the children voted on their favorite smells. He had one that smelled like a skunk, but not very many chose that as their favorite!

Ms. Tsai liked stickers, too. She also liked cows and had a cow stuffed animal on her desk. His name was Moo Goo Gai Pan. That is the name of a Chinese food, but she named him that because of the Moo part!

One Friday Ms. Tsai said she had new stickers that were sure to be her favorites. The children all tried to guess what they were. Pedro guessed rainbows, and Jackie Sue thought they were bears, but Ms. Tsai said nobody had guessed right so far. Can you guess what they were? They were four different kinds of cow stickers! After the children each picked one, there was one cow sticker left. Pedro said, "Let's put it on Moo Goo Gai Pan!" and that's exactly what they did.

Discussion Questions

1. What did the children in Ms. Tsai's class love? (*stickers*)
2. What were some of the kinds they collected? (*dinosaurs, scratch and sniff*)
3. What was Ms. Tsai's new kind of sticker? (*cow stickers*)
4. What is your favorite kind of sticker?

5 READINESS ACTIVITY

Tell How Numbers Are Used in the Real World

Story

Brian and Jackie Sue have been saving their allowances for stickers. Jackie Sue can't decide if she should buy the large stickers or the small ones. The large stickers cost more so she couldn't buy as many, but she thinks they're more fun.

Questions

1. How much do the large stickers cost? (*10¢*)
2. How much do the small stickers cost? (*5¢*)
3. Why do you think some stickers cost more than others? (*they are larger and fancier*)
4. Can you find the cash register? Why do cash registers have a window that shows the price? (*so people can see what their purchase will cost*) What price does it show? (*.10 or 10¢*)
5. What do the numbers on the clock tell you? (*what time it is*) Do people sometimes carry a kind of clock with them? (*yes, a watch*) Why might it be useful to know what time it was?

TEACHING ACTIONS

1. Read and discuss the story.
2. Using the sample questions given, discuss how numbers are used in the picture.
3. (*optional*) Extend discussion to children's own experiences with shopping, allowances, watches, and other personal numbers.

6 READINESS ACTIVITY

Answer Questions About a Story

Story A

Melody went to the Science Museum to see the new dinosaur exhibit. She bought 3 large dinosaur stickers and 6 small ones. When she gets home, she will put them in her sticker book.

Questions for Story A

1. Where did Melody go? (*to the Science Museum*)
2. How many large stickers did she buy? (*3*) How many small ones? (*6*) How many stickers in all? (*9*)
3. What is she going to do with them? (*put them in her sticker book*)

Story B

At the Science Museum, Melody also learned about wolves. There were 2 adult wolves and 3 baby wolf pups standing in some grass and trees. The wolves looked so real that at first she thought they were alive!

Questions for Story B

1. Melody saw the dinosaur exhibit. What else did she learn about at the museum? (*wolves*)
2. How many adult wolves did she see? (*2*) How many baby pups? (*3*) How many wolves in all? (*5*)
3. Why did she think they were alive? (*they looked real*)
4. Have you ever been to a museum? What was the most interesting thing you saw there?

TEACHING ACTIONS

1. Read Story A.
2. Ask the questions for Story A.
3. Repeat for Story B.

7 PROCESS PROBLEM

Jackie Sue likes to make designs with her stickers. For this page of her sticker book, she pasted a row of five stickers across the bottom of the page. (Have children put five tiles on the bottom row.) For the row above the bottom, she pasted four stickers. (Have children put a row of four tiles above the row of five tiles. Make sure the rows line up at the left.) The next row up she put three stickers. (Have children place a row of three tiles.) How many rows in all did she have when she was done?

MATERIALS

1" color tiles or 1" construction-paper squares (at least 15 per child or pair)

Understanding the Problem

- What is Jackie Sue doing with her stickers? (*putting them in her book in a design*)
- With what row did she start? (*the bottom row*)
- What row did she paste in next? (*the next row up*)
- What are we trying to find? (*how many rows she had in all when she was done*)

Solving the Problem

- How is this second row up (point to the row with four stickers) alike or different from this bottom row (point to the bottom row)? (*the bottom row has five stickers; the second row up has four*)
- Repeat the first question for the second and third rows up. (*four stickers vs. three stickers*)
- What will Jackie Sue do next? (*paste another row above the row of three*)
- How many stickers the next row will have? (*2*) How do you know? (*there's one fewer sticker in each new row*) Have children place a row with two tiles above the row with three tiles.
- Can she paste another row? (*yes, it will have one sticker*) Have children place one tile for this row.
- What will she do now? (*stop, because a row can't have less than one sticker*)

STRATEGY ASSESSMENT IDEAS

Listen and watch as children work to see if they

- describe a pattern formed by the decreasing number of tiles in each subsequent row
- extend the pattern correctly with tiles
- use the pattern to arrive at a correct solution to the problem

7 Set 2 Favorite Stickers

PROBLEM-SOLVING EXPERIENCES IN MATHEMATICS K LEVEL

- Do we need to do anything else to solve the problem? (*we need to count how many rows there are—5*)

Solution

Look for a Pattern

Pattern: (from the bottom row) five stickers, four stickers, three stickers, two stickers, one sticker

There are five rows of stickers.

Problem Extensions

1. Have the children trace around the tiles in the bottom row and write the numbers 1 through 5 in the tiles. Repeat for the other rows. Have them look for number patterns.
2. Have children make their own sticker patterns with tiles and have partners extend them.

8 PROCESS PROBLEM

Ms. Tsai puts a sticker on a chart each day so the children will know what they will be doing at work time. On day 1 she put on a square sticker to show they would be working at tables. (Have children put a square on the square for day 1.) On day 2 she put on a round sticker for working at stations. (Have children put a counter on day 2's circle.) Day 3 was tables, and day 4 was stations. (Have children place the next square and counter.) If Ms. Tsai keeps to the same schedule, where will the children be working on day 5? Where will they be working on day 6 and day 7?

MATERIALS

1" counters or pattern block triangles, and pattern block squares (at least 4 of each per child or pair)

Understanding the Problem

- How does Ms. Tsai let the children know where they will be working? (*by putting stickers on a chart*)
- What kind of sticker means working at tables? (*a square one*) Working at stations? (*a round one*)
- What are the words labeling the spaces for stickers? (*the days of the week*)
- What are we trying to find? (*where the children will be working on day 5, day 6, and day 7*)

Ways to Solve the Problem

- How is this second sticker (point to the day 2 round sticker) alike or different from this first sticker (point to the day 1 square sticker)? (*day 2's sticker is round and day 1's is square*)
- Repeat for day 3 and day 2, and day 4 and day 3. (*square and round, round and square*)
- What shape will the next sticker be? (*square*) How can you tell? (*the stickers form a square-round-square-round pattern*) Have children place manipulatives to continue the pattern.
- Where will the children be working on day 5? (*at tables, because the sticker is square*)
- Where will they be working on day 6 and day 7? (*stations, then tables*)

▶ *turn the page*

STRATEGY ASSESSMENT IDEAS

Listen and watch as children work to see if they

- describe a pattern formed by the shapes of the stickers
- extend the pattern correctly with squares and triangles
- use the pattern to arrive at a correct solution to the problem

Solving the Problem

Look for a Pattern

Pattern: square (*tables*), round (*stations*), square (*tables*), round (*stations*)

Day 5 is tables, day 6 is stations, day 7 is tables. Have children draw in the correct shapes.

Related Problem: 7

Problem Extensions

1. Have children place manipulatives in the following sequence: square, round, square, square, square, round. Tell them Ms. Tsai made a mistake. Which one do you think is wrong? (*the third square*)
2. Have children decide on a shape for free time. Use the following pattern: tables, stations, free time, tables, stations, free time, tables. What will the children be doing on day 8? (*stations*)

A Trip to the Zoo

Ethan and Joleen think they are lucky because they live very close to the city zoo. Ethan likes the monkeys the best, but Joleen's favorites are the dolphins. Last time they visited the zoo they saw the monkeys first. This time it's Joleen's turn to choose, so the family is going first to the dolphin pool.

Joleen likes to see the dolphin trainers hold fish out over the water. The dolphins swim around the pool very fast, then jump all the way out of the water to grab the fish with their mouths. When the dolphins land in the water, they make a big splash! On hot days, Joleen likes to stand close to the edge of the pool and get splashed.

After they see the dolphins, Joleen and Ethan will go to the monkey area. On the way they will pass the zebras and the tigers. Their family has a striped cat at home named Tiger, but the zoo's tigers are much bigger than their cat. Joleen is glad that their cat isn't that big! Ethan thinks it's funny that the zoo keepers put two striped animals right next to each other. This visit he's going to see if any other animals that are alike are next to each other.

Discussion Questions

1. Why do Ethan and Joleen think they are lucky? (*they live close to the zoo*)
2. What are Joleen's favorite animals? (*the dolphins*)
3. Why does Joleen want to stand near the dolphins on hot days? (*so she can get splashed*)
4. What is your favorite animal at the zoo?

9 READINESS ACTIVITY

Tell a Number Story

Story

Ethan and Joleen saw a bird show at the zoo. Some of the birds were brightly colored parrots, and 3 of them could talk. One of them said, "Hello. Thank you for coming!"

Task 1

Tell a story about the zoo that has the number 4 in it.

Task 2

Tell a story about the zoo that has the number 7 in it.

TEACHING ACTIONS

1. Read and discuss the story.
2. Give Task 1 to the children and model it. Solicit a variety of stories. Encourage children to make up challenging stories.
3. Repeat for Task 2.

10 READINESS ACTIVITY

Answer Questions About a Story

Story A

At the zoo store, Ethan and Joleen's father told them they could each pick out 3 animal cards. He said if they each picked different cards, he could teach them a mystery game to play with the animals on the cards.

Questions for Story A

1. What did Ethan and Joleen's father tell them they could buy? (*animal cards*)
2. How many cards did each buy? (3)
3. What is their father going to do? (*teach them a mystery game to play with the cards*)

Story B

When the family got to the monkey area, Ethan saw 6 adult monkeys. Two of them had baby monkeys riding on their backs. The babies were holding onto their mother's necks so they wouldn't fall.

Questions for Story B

1. How many adult monkeys did Ethan see? (6)
2. How many had babies? (2)
3. What were the baby monkeys doing? (*holding onto their mother's necks*)
4. Have you ever seen monkeys? What is your favorite thing to see them do?

TEACHING ACTIONS

1. Read Story A.
2. Ask the questions for Story A.
3. Repeat for Story B.

11 PROCESS PROBLEM

At the entrance to the zoo, a sign told the times when each of the animals would be fed. Ethan and Joleen's mother wrote down the times, and their family saw all their favorites getting fed. The dolphins ate fish. The monkeys ate monkey pellets. The rabbits didn't eat the hay. What did the elephants eat?

MATERIALS

1" counters or 1" construction-paper squares or pattern blocks (2 each of 3 colors per child); crayons

Understanding the Problem

- What were Ethan and Joleen's family doing? (*watching animals get fed*)
- Who ate pellets? (*monkeys*)
- Who ate fish? (*dolphins*)
- What didn't the rabbits eat? (*hay*)
- What are we trying to find? (*what the elephants ate*)

Solving the Problem

- Help children identify each of the animals pictured (*dolphin, monkey, rabbit, elephant*)
- Help children identify each of the foods pictured (*hay, monkey pellets, fish, lettuce*)
- Can you find the food the dolphins ate? Place same-color counters on the dolphin and its food. (*fish*)
- Can you find the food the monkeys ate? How could you show they belong together? (*put matching counters of a new color on the monkeys and the pellets*)
- If the rabbits didn't eat the hay, what could they eat? (*lettuce*) How could you show they belong together? (*put matching counters of a new color on the rabbits and the lettuce*)
- Is there an animal left? (*yes, the elephant*)
- Is there a food left? (*yes, hay*)
- What did the elephant eat? (*hay*) Have children draw lines connecting each animal with its food.

Solution

Use Logical Reasoning

The elephant and the hay are the only things not matched. The elephant ate hay.

Problem Extensions

1. Suppose the elephants ate the lettuce. What would the rabbits eat? (*hay*)
2. Have 4 children secretly tell you a favorite food. Write the children's names on the board. Sketch the foods in another column, not necessarily next to the name. Name three of the children and their foods. Discuss with the class how to find the last child's food. Have children come in turn to draw the connecting lines.

STRATEGY ASSESSMENT IDEAS

Listen and watch as children work to see if they

- correctly use counters or draw lines to match animals with their food
- correctly use all statements given in the problem
- conclude that the remaining animal and remaining food go together

12 PROCESS PROBLEM

Ethan and Joleen's father lay the animal cards on the table and taught them the mystery game. He secretly picked one of the animals to be his mystery animal. He gave 3 clues about his animal. Ethan and Joleen had to guess which animal he had picked. You can play, too. Look at the picture, and use the 3 clues to find out the mystery animal.

MATERIALS

1" counters, 1" tiles, or 1" construction-paper squares (5 per child); crayons

Understanding the Problem

- What did Ethan and Joleen's father teach them? (*a mystery game*)
- What did he secretly pick? (*one of the animals*)
- What is a clue? (*a hint*) What is a mystery? (*something we don't know*)
- What are we trying to find? (*what his mystery animal is*)

Solving the Problem

- Help children identify each of the animals pictured. (*elephant, snake, zebra, fish, turtle, kangaroo*)
- Read the first clue to the children. Which animals walk on four legs? (*elephant, zebra, turtle*)
- Are there animals who can't be the mystery animal? (*the fish, kangaroo, and snake do not walk on four legs*) Place a counter on the fish, kangaroo, and snake to show they are "out."
- Read the second clue to the children. Of the 3 animals left, which have small ears? (*zebra, turtle*)
- Are there animals who can't be the mystery animal? (*elephant*) Place a counter on the elephant.
- Read the third clue to the children. Who has stripes? (*zebra*)
- Are there animals left who aren't the mystery animal? (*turtle*) Place a counter on the turtle.
- What is the mystery animal? (*zebra*)

▶ *turn the page*

STRATEGY ASSESSMENT IDEAS

Listen and watch as children work to see if they

- correctly use counters to show animals that have been eliminated
- correctly use all statements given in the problem
- conclude that the remaining animal is the mystery animal

Solution

Use Logical Reasoning

Clue 1 tells us that the mystery animal is the elephant, zebra, or turtle. The fish, kangaroo, and snake are eliminated.

Clue 2 tells us it is the zebra or turtle. The elephant is eliminated.

Clue 3 tells us it is the zebra. The turtle is eliminated.

Related Problem: 11

Problem Extensions

1. Give a different set of clues:
   a. I usually swim or hop. (*eliminates elephant, zebra, and snake*)
   b. I live in the water. (*eliminates kangaroo*)
   c. I have a hard shell. (*eliminates fish; the turtle is the mystery animal*)
2. Secretly pick one of the animals. Have children ask yes-or-no questions. For example: Does it live in the water? Have children place counters on their pictures as animals are eliminated.

Bath Toys and Water

Drew loved to play with toys in the bathtub and sink. He had many floating toys like boats, ducks, dinosaurs, and people. Sometimes he put the people on the boats and pretended the dinosaurs were swimming.

Drew also had a collection of plastic bottles, margarine tubs and milk cartons. He liked to hold a measuring cup full of water high above the containers. Then he would pour water into them until they sank.

One day Drew decided to do a science experiment. He put a margarine tub in the sink and guessed how much water it would hold. He ran just that much water into the measuring cup. He was just starting to pour the water from the measuring cup into the margarine tub when his mother called him. Drew turned to talk to her, but he forgot to stop pouring! Water went all over the floor. Drew decided his next experiment would be to see how many towels it took to clean up the mess!

Discussion Questions

1. What did Drew like to play with? (*toys in the bathtub or sink*)
2. How did he sink his plastic containers? (*by pouring water into them*)
3. What happened when he was pouring water from the measuring cup? (*water went all over the floor*)
4. Have you ever spilled water?

13 READINESS ACTIVITY

Answer Questions About a Story

Story A

Veejay collects ducks to float in the bathtub. He has 10 ducks. He sorted the 5 yellow ducks into one group and the 4 orange ducks into another. He had a pink one left over.

Questions for Story A

1. What did Veejay collect? (*ducks to float in the bathtub*)
2. Do you know the colors of the ducks? (*yellow, orange, pink*)
3. How many groups of ducks had more than one duck in them? (*2*)

Story B

Veejay put the 5 yellow ducks in a line. The line just fit across the bathtub. The line of 4 orange ducks didn't reach to the other side.

Questions for Story B

Note: You may want to line up manipulatives in a graph fashion to help with this story.

1. What did Veejay do with the yellow ducks? (*he put them in a line*)
2. How far did the line of yellow ducks reach? (*across the bathtub*)
3. Were the lines of ducks equal in length? (*no*)
4. How many more yellow ducks were there than orange ducks? (*1*)

TEACHING ACTIONS

1. Read Story A.
2. Ask the questions for Story A.
3. Repeat for Story B.

14 READINESS ACTIVITY

Visualize a Story

Story A

Miranda and Sara were using straws to blow boats across the kitchen sink. Miranda put the end of her straw close to her boat and blew through the straw. The boat sailed right across the water. Sara had the end of her straw pointed at the water instead of at her boat, so her boat didn't go very far.

Story B

Some of the kindergartners brought dinosaurs that floated to school. They filled the sink with water. Arman called his dinosaurs 1 and 2. Martha called hers 3 and 4. Chantal called hers 5 and 6. The children held them under the water and tried to let go of them in turn so they would pop up in order: 1, 2, 3, 4, 5, 6!

TEACHING ACTIONS

1. Have children close their eyes.
2. Tell children to picture in their minds the story you will read.
3. Read Story A.
4. Ask children to describe what they visualized. Ask specific questions such as: What does the boat look like?
5. Read Story B.
6. Ask children to describe what they visualized. Ask specific questions such as: What colors were the dinosaurs? What did the children look like?

15 PROCESS PROBLEM

Drew was pretending his boats were sailing down a river. (Have children place a plastic link on each boat.) He drew a line on a piece of paper to show the river. Then he put all his boats in a line on the river. How many of his boats do you think he could fit on the river?

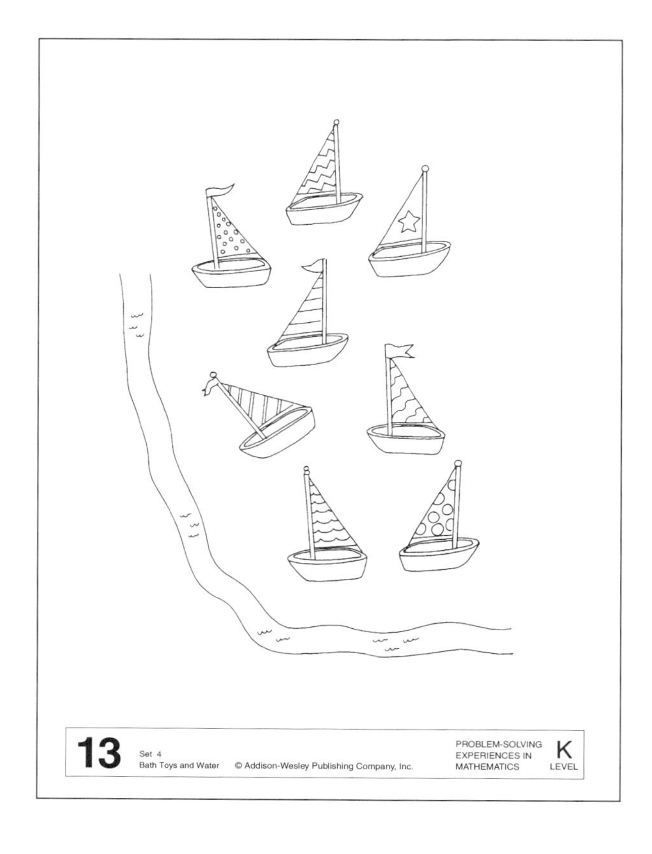

MATERIALS

$1\frac{3}{4}$" plastic links (8 per child or pair)

Understanding the Problem

- What was Drew pretending? (*that his boats were sailing down a river*)
- What are we using for boats? (*plastic links*)
- How is he showing the river? (*with a line on his paper*)
- What are we trying to find? (*how many boats will fit on the river*)

Solving the Problem

- How many boats does Drew have? (*8*)
- How many boats do you think will fit on the river? Elicit guesses.
- How can you tell if your guess is right? (*move that many links onto the line*) Have children move links onto the line.
- If your guess wasn't right, can you make a new guess? How can you check? (*add more boats or take some away*)
- How many boats can fit on the river? (*5*)

Solution

Guess and Check

There will be 5 boats on the river.

Related Problems: 4, 3

Problem Extensions

1. Repeat the activity using materials of other lengths, such as paper clips. Do children understand that fewer large boats or more small boats can fit on the river?
2. Give a different pattern block shape to each small group. Have children estimate how many of their block it will take to make a line the whole length of the paper. Do they understand that it will take more of the smaller blocks? Compare results between groups.

STRATEGY ASSESSMENT IDEAS

Listen and watch as children work to see if they

- make a reasonable first guess (a range of 3–8 boats is reasonable)
- make a second guess using what they learned from checking with the links
- use links to check their guesses

16 PROCESS PROBLEM

Chantal made a parade of six boats. Some of the boats were red and some were blue. There was the same number of red boats as blue boats. How many boats of each color were there?

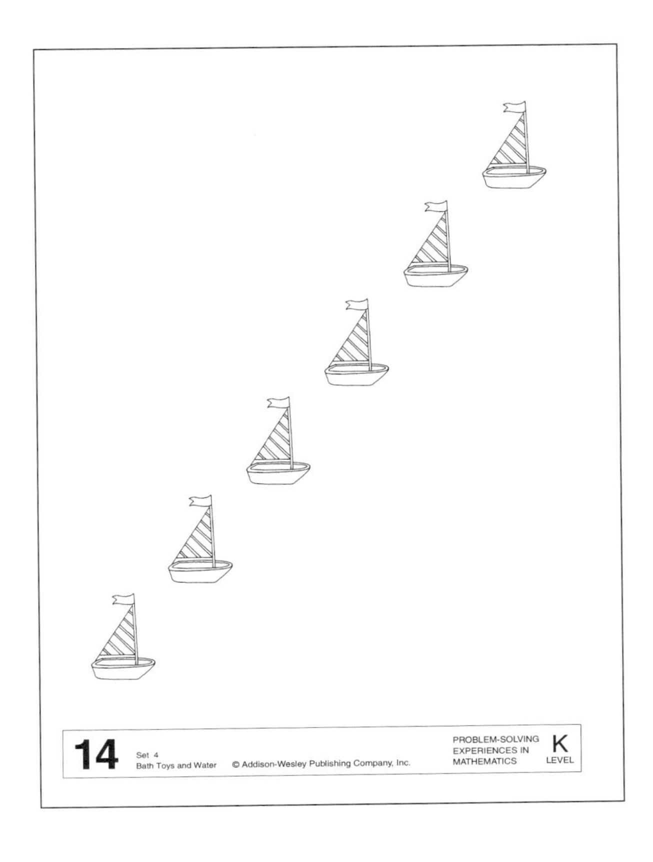

MATERIALS

red and blue 1" counters or 1" construction-paper squares (6 of each per child or pair)

Understanding the Problem

- What did Chantal do with her boats? (*made a parade of them; put them in a line*)
- What colors were the boats? (*red and blue*)
- How many boats were in the parade? (6)
- Were there more red boats, more blue boats, or the same number of each? (*the same number of each*)
- What are we trying to find? (*how many there were of each color of boat*)

Solving the Problem

- Elicit guesses of numbers of red and blue boats.
- Could we have all blue boats? (*no, there are red and blue boats*)
- Could we have 1 red boat and 1 blue boat? Have children put 1 red counter on a boat and 1 blue counter on another boat. (*no, there are 6 boats*)
- Do you need to change your guess? How can you check? (*add red and blue counters to make the numbers guessed*)
- How will you know you are right? (*there will be the same number or red and blue boats; there will be 6 boats*) How many red and blue boats were there? (*3 red, 3 blue*)

Solution

Guess and Check

3 + 3 = 6

There are 3 blue boats and 3 red boats.

STRATEGY ASSESSMENT IDEAS

Listen and watch as children work to see if they

- make a first guess that indicates an understanding of the problem (an equal number of red and blue boats, a total of 6)
- make a second guess using what they learned from checking with the counters
- use counters to check their guesses

Related Problems: 15, 4, 3

Problem Extensions

1. Repeat the activity using a parade of 8 or 10 boats. (*4 boats of each color; 5 boats of each color*)
2. Use three colors and a parade length of 9 boats. (*3 boats of each color*)

A Circus Surprise

Last night when Josette and her family watched the news on television, she saw pictures of circus elephants walking down a ramp from a circus train. Now the day had come at last—they were going to see the circus in person!

Josette's seat was on an aisle. Her big sister, Marie, sat next to her. Finally, the lights went out. A spotlight shone in the middle of the ring, on a man with a tall top hat. Josette knew that this was the ringmaster who would make all the announcements. "Ladies and gentlemen," he said, "here comes the circus parade!"

There were dogs, people standing on white horses, and lions and tigers. At the end came the elephants. But . . . there were no clowns. Suddenly there was a very loud "TOOT! TOOT!" next to Josette's ear. She was so surprised she almost jumped out of her seat! The clowns were coming down the aisles and one had honked his horn right next to her. He shook her hand and handed her a silver and green balloon! Josette saw many surprising things at the circus that day, but thought the clown and balloon were the best surprise of all.

Discussion Questions

1. What did Josette see on the news? (*circus elephants getting off a circus train*)
2. Where was Josette's seat? (*on the aisle*)
3. What kinds of acts were in the parade? (*dogs, people on horses, lions and tigers, elephants*)
4. Why was Josette surprised? (*the clown honked his horn next to her ear*)
5. If you worked in a circus, what would you like to do?
6. Were you ever surprised? What do you think was your favorite surprise?

17 READINESS ACTIVITY

Tell a Number Story

Story

When it was time for the elephant act, the elephants came into the ring in a line. Two of the elephants had people sitting on their backs, but the other 3 didn't have riders.

Task 1

Tell a story about the circus that has the number 5 in it.

Task 2

Tell a story about the circus that has the number 1 in it.

TEACHING ACTIONS

1. Read and discuss the story.
2. Give Task 1 to the children and model it. Solicit a variety of stories. Encourage children to make up challenging stories.
3. Repeat for Task 2.

18 READINESS ACTIVITY

Visualize a Story

Story A

Josette's big sister, Marie, goes to gymnastics class and could hardly wait to see the acrobats. First the acrobats set up a small trampoline. Then one of them ran across the ring, jumped on the trampoline, and flew into the air. She did two somersaults in the air before landing on the mat! Marie wished she could do that!

Story B

At the end of their act, 3 of the acrobats stood in a line. Two more climbed onto their shoulders. Then the smallest acrobat climbed to the very top of the pyramid and waved a flag!

TEACHING ACTIONS

1. Have children close their eyes.
2. Tell children to picture in their minds the story you will read.
3. Read Story A.
4. Ask children to describe what they visualized. Ask specific questions such as: What was the acrobat wearing? What smells could Josette and Marie smell?
5. Read Story B.
6. Ask children to describe what they visualized. Ask specific questions such as: Were the acrobats men or women? What color was the flag?

19 PROCESS PROBLEM

After the acrobat act, a very small car came out and drove to the center of the ring. The driver got out, and guess who it was—the clown who had surprised Josette! He was holding a silver and green balloon just like hers. Then clown after clown climbed out of the tiny car. Soon there were 6 clowns standing in the ring. Josette couldn't see what the last 2 looked like. Josette noticed her clown had a red hat and the next clown, with the dog, had a yellow hat. The third clown, with a feather, had a red hat. The fourth clown, with a purse, had a yellow hat. (Have children place counters on the appropriate hats.) Marie said, "Look at the colors of the hats. They make a pattern." What color hats were the last 2 clowns wearing?

MATERIALS

red and yellow 1" counters, tiles, or 1" construction-paper squares (4 of each per child or pair); red and yellow crayons

Understanding the Problem

- What were the clowns doing? (*getting out of a car*)
- Which clowns couldn't Josette see? (*the last 2*)
- Whose hat colors could she see? (*the first 4 clowns*)
- Why do we know there's a pattern? (*Marie said the hat colors made a pattern*)
- What do we need to find? (*the color of the hats of the last 2 clowns*)

Solving the Problem

- What color is this hat? (point to the clown with the balloon) (*red*)
- What color is the next hat in line? (point to the clown with the dog) (*yellow*)
- What color is this hat? (point to the clown with a feather) (*red*)
- What color is the next hat in line? (point to the fourth clown, with the purse) (*yellow*)
- Can you figure out the last two hats? Put counters on the hats and check to see if you have a color pattern for all the clowns' hats. Then color the hats the color of the counters.

Solution

Look for a Pattern

Pattern: (hat colors from left to right) red, yellow, red, yellow, red, yellow

Clown with balloon: red; clown with dog: yellow; clown with feather: red; clown with purse: yellow; firefighter: red; clown with trumpet: yellow.

Related Problems: 8, 7

Problem Extension

Have children look for other patterns in the line of clowns: buttons/no buttons; small shoes/large shoes.

STRATEGY ASSESSMENT IDEAS

Listen and watch as children work to see if they

- describe a pattern formed by the colors of the hats
- extend the pattern correctly with yellow and red counters
- use the pattern to arrive at a correct solution to the problem

20 PROCESS PROBLEM

At the very end of the circus, the performers had another parade. The 10 band people marched in pairs. As each pair came into the spotlight, Josette could see their instruments. The first 2 were playing drums, and the second 2 were playing trumpets. The third pair was playing drums. (Have children place counters on drums, and links on trumpets.) Josette said, "I know what instruments the next pairs will be playing, and I think there will be 6 drummers in the whole band." What instruments will the last 2 pairs of band people be playing? Was Josette right about the number of drums?

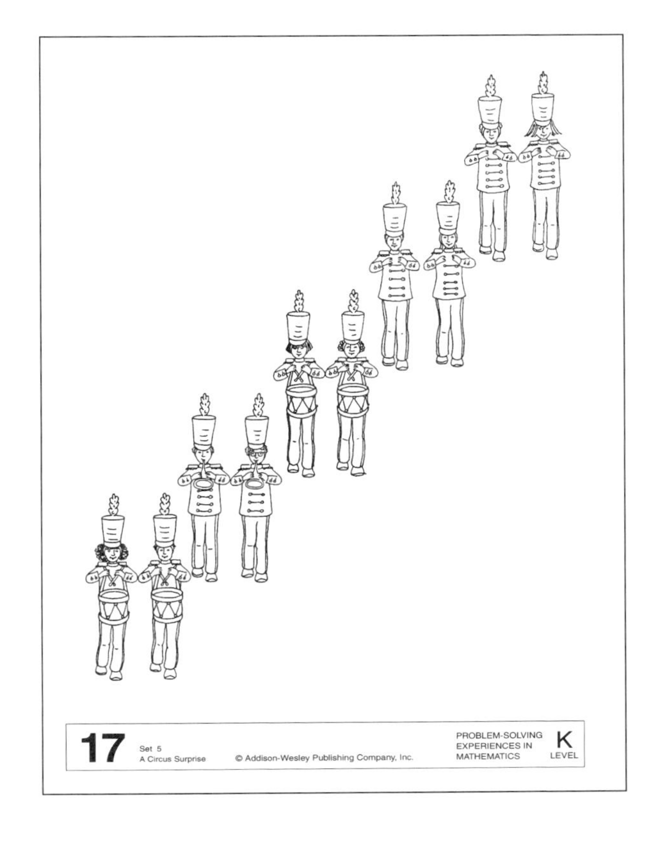

MATERIALS

1" counters or 1" construction-paper squares, and plastic links (at least 6 of each per child or pair); crayons

Understanding the Problem

- What was the band doing? (*playing in a parade at the end of the circus*)
- How were they marching? (*in pairs, two at a time*)
- What instruments were the band people playing? (*drums and trumpets*)
- What are we trying to find? (*what instruments the last 2 pairs of band people were playing; how many people were playing drums in the whole band*)

Solving the Problem

- How is this pair (point to the first pair) different from this pair (point to the second pair)? (*the first pair is playing drums, the second pair is playing trumpets*)
- Repeat for the second and third pair. (*trumpets and drums*)
- What do you think the next pair will be playing? Why? (*the drums and trumpets form a pattern*) Put counters or links on the next pair and check. (*links—trumpets, counters—drums*) Color in the missing instruments.
- What else did we need to know? (*how many drums were in the band and whether Josette was right*)

▶ *turn the page*

STRATEGY ASSESSMENT IDEAS

Listen and watch as children work to see if they

- describe that the drums and trumpets form a pattern
- extend the pattern correctly with counters and links
- use the pattern to arrive at a correct solution to the problem

Solution

Look for a Pattern

Pattern: drums (counters), trumpets (links), drums, trumpets, drums, trumpets

The last 2 clowns are playing drums. There are 6 drums in the band; Josette was right.

Related Problems: 19, 8, 7

Problem Extensions

1. Have children place manipulatives on the hats of the first 4 pairs to form a repeating pattern of their choice. Have a partner complete the pattern.
2. If the band had two more players, what instruments would they be playing? (*trumpets*)

A Cooking Lesson

Ms. Bohlig's kindergartners have been learning about different jobs people do. Last week, Dau's mother, Dr. Tran, talked to the class about being a doctor. She brought a stethoscope and the children listened to the sounds their hearts made. Tony said his sounded like "lup dub, lup dub."

This week Frankie's father is visiting. He is a cook at an Italian restaurant, and he is called Chef Morelli. He brought dough, peppers, tomato sauce, mushrooms, and olives. He also brought a large knife, a wooden cutting board, and a flat, round pan. Can you guess what he was going to make? Pizza!

First he stretched the lump of dough into a circle. He twirled it around in the air until it was much larger and thinner. When he had turned up the edges, he spread the tomato sauce on it. Then he cut up the peppers and mushrooms. The children were amazed to see how fast he could cut! Chef Morelli said he'd had plenty of practice. He reminded them to be very careful when they used knives.

When Chef Morelli was done, Mr. Beardsley, the school cook, let him cook the pizza in the school oven. It smelled so good! Everyone had a piece when it was done. They thought it was the best they'd ever had!

Discussion Questions

1. What is Ms. Bohlig's class learning about? (*different jobs people do*)
2. Who has visited the class so far? (*Dau's mother, Dr. Tran, and Frankie's father, Chef Morelli*)
3. What kind of job does Chef Morelli have? (*he's a cook*)
4. How did he make the dough look like a pizza? (*he stretched it into a circle, then twirled it in the air*)
5. What did he put on the pizza? (*tomato sauce, peppers, mushrooms, and olives*)
6. Have you ever made pizza? If you were to make one, what would you put on it?

21 READINESS ACTIVITY

Visualize a Story

Story A

Cory told his mother about Chef Morelli's visit, and they decided to bake cookies together. Cory's mother showed him how to measure 1 cup of flour. They put this in a bowl. Then they measured half a stick of margarine. They cut it into little pieces, and put the pieces in the bowl, too.

Story B

They needed 2 eggs for the cookies. Cory hit the first egg against the side of the bowl to crack it. At first he didn't hit it hard enough and nothing happened. Then he hit it so hard pieces of eggshell flew into the bowl along with the egg! His mother said the same thing happened to her sometimes.

TEACHING ACTIONS

1. Have children close their eyes.
2. Tell children to picture in their minds the story you will read.
3. Read Story A.
4. Ask children to describe what they visualized. Ask specific questions such as: What kind of cookies are they making?
5. Read Story B.
6. Ask children to describe what they visualized. Ask specific questions such as: What does Cory look like?

22 CONCEPT/SKILL ACTIVITY

Complete a Picture to Show a Story

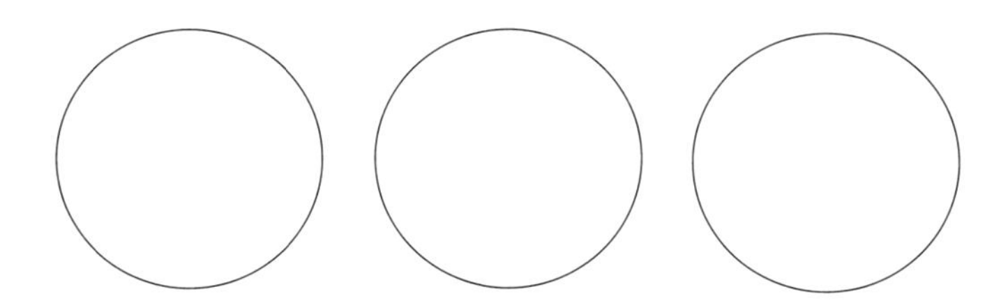

MATERIALS

drawing paper; 1" counters or 1" construction-paper squares (3 each per child of orange, green, and white for Story A; 3 each of red and yellow for Story B); orange, green, white, red, and yellow crayons

Story A

Chef Morelli was preparing dinners for 3 people. Two wanted orange sweet potatoes and one wanted spinach. They all wanted a roll.

Story B

The person who wanted spinach also wanted red beets. One of the other people wanted beets, and the last person wanted yellow corn.

TEACHING ACTIONS

1. Draw the circles shown on the board, and have the children copy the picture. Children's circles should be about 4" in diameter. Tell the children these are dinner plates.
2. Read Story A to the children. Have them place counters on the plates to show food: orange for sweet potatoes, green for spinach, and white for rolls. Then have them draw small colored circles on the plates to show the food, and remove the counters.
3. Have several children share their pictures.
4. Read Story B to the children. Have them place counters on the plates: red for beets and yellow for corn. (plates will contain sweet potatoes, a roll, and beets; sweet potatoes, a roll, and corn; spinach, a roll, and beets) Have them draw the food and remove the counters.
5. Have several children share their pictures. Compare different arrangements of food. Are there 3 correct arrangements of food even though they may be in different places?
6. (*optional*) Have the children tell how many servings there are of each type of food. (*2 of sweet potatoes, 1 of spinach, 3 rolls, 2 of beets, 1 of corn*) How many different foods? (*5*) How many portions in all? (*9*)

23 PROCESS PROBLEM

Cory, his mother, and his friend Dau each made a cookie from Cory's cookie dough. Cory made the largest cookie. His mother's cookie was smaller than Dau's. Who made the middle-size cookie?

MATERIALS

1" counters, 1" tiles, or 1" construction-paper squares (2 each of 2 colors per child or pair); crayons

Understanding the Problem

- How many cookies did each person make? (*1*)
- Were they all the same size? (*no*)
- Who made the largest cookie? (*Cory*)
- Was Cory's mother's cookie larger or smaller than Dau's? (*smaller*)
- What are we trying to find? (*who made the middle-size cookie*)

Solving the Problem

- Can you put matching counters on Cory and his cookie? (*the largest one*)
- Could Dau have made the smallest one? (*no, Cory's mother's is smaller than Dau's*)
- Choose a different color and put matching counters on the smallest cookie and the person who made it. (*Cory's mother*)
- Who made the middle-size cookie? (*Dau*) Draw lines between each person and his or her cookie.

Solution

Use Logical Reasoning

Dau made the middle-size cookie.

Related Problems: 12, 11

Problem Extensions

1. If Cory made the smallest cookie instead of the largest cookie, who made the largest? (*Dau*)
2. Suppose Cory's mother made a middle-size cookie, and Dau's was smaller than Cory's. Who made the smallest cookie? (*Dau*)

STRATEGY ASSESSMENT IDEAS

Listen and watch as children work to see if they

- use counters to match people with their cookie size
- correctly use all statements given in the problem
- conclude that the remaining person and the middle-size cookie go together

24 PROCESS PROBLEM

Cory's father loves to bake bread. He showed Cory how to make the dough, then how to shape it into a loaf. Cory, his mother, and his father each made a loaf of bread. One loaf was very short. Another was longer. The last loaf was very long. Cory's father made the longest loaf. Cory didn't make the middle-size loaf. Whose loaf was the shortest?

MATERIALS

1" counters, 1" tiles, or 1" construction-paper squares (2 each of 2 colors per child or pair); crayons

Understanding the Problem

- What were Cory, his father, and his mother doing? (*shaping loaves of bread*)
- Were they all the same length? (*no*)
- Who made the longest loaf? (*Cory's father*)
- Did Cory make the middle-size loaf? (*no*)
- What are we trying to find? (*whose loaf was the shortest*)

Solving the Problem

- Can you put the matching counters by the longest loaf and the person who made it? (*Cory's father*)
- Who is left? (*Cory and his mother*) If Cory didn't make the middle-size loaf, who did? (*Cory's mother*) Can you choose a different color and show this with counters?
- Who is left? (*Cory*) What size loaf did he make? (*the shortest*)
- Have the children draw lines between each person and the loaf he or she made.

Solution

Use Logical Reasoning

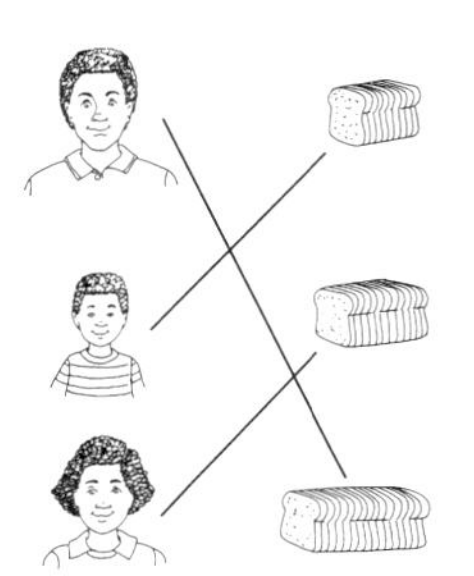

Cory's loaf was the shortest.

20 Set 6 A Cooking Lesson

PROBLEM-SOLVING EXPERIENCES IN MATHEMATICS K LEVEL

Related Problems: 23, 12, 11

Problem Extension

Suppose Cory's mother made the longest loaf, and his father didn't make the shortest one. Who made the middle-size one? (*his father*)

STRATEGY ASSESSMENT IDEAS

Listen and watch as children work to see if they

- use counters to match people with their loaf
- correctly use all conditions given in the problem
- conclude that the remaining person and the shortest loaf go together

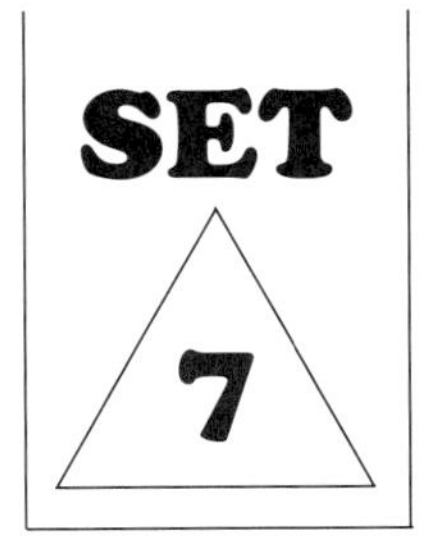

Valentine's Day

Beth had invited 5 of her friends for a Valentine's Day party. At the store, Beth chose plain pink paper plates and cups. When she got home, she drew hearts on them with a red marker. Her sister, Jen, helped her spell her friends' names—everyone had a special plate and cup.

At the party, Beth's brother, Jeff, came in with 2 paper bags. He said there were surprises in each bag. He would give the children clues for the first bag. If they guessed how many things were in the first bag, they got what was in that bag. The first clue was: The number is less than the number of toes on 2 feet. Crystal said that meant there were fewer than 10 things in the bag. The second clue was: The number is more than the number of fingers on one hand. Joey said, "The number is between 5 and 10!"

Marco guessed 8, but Jeff said that was too many. Kia guessed 6 and Jeff said, "Yes! You guessed it!" There were 6 cookies in the bag, just enough for each child at the party. And . . in the other bag were tubes of icing so the children could decorate their own cookies! Joan said, "This isn't just a Valentine's Day party—it's a surprise party, too!"

Discussion Questions

1. What did Beth invite her friends for? (*a Valentine's Day party*)
2. How did Beth decorate the paper plates and cups? (*with hearts and names, using a red marker*)
3. How many bags did Beth's brother Jeff have? (2)
4. What clues did he give? (*the number was less than the number of toes on 2 feet, more than the number of fingers on 1 hand*)
5. What was the second surprise? (*tubes of icing*)
6. Have you ever played a guessing game? What was it like?

25 READINESS ACTIVITY

Retell a Number Story

Story

Mai was passing out treats to the class. She passed out 4 heart-shaped cookies and 5 that were shaped like stars.

Task 1

Retell the story with a different number of cookies.

Task 2

Retell the story with stickers instead of cookies.

TEACHING ACTIONS

1. Read and discuss the story.
2. Give Task 1 to the children. Model it, then solicit stories with different numbers of cookies in them.
3. Repeat for Task 2.

26 CONCEPT/SKILL ACTIVITY

Choose Missing Data

MATERIALS

1" counters or 1" construction-paper squares for Problem A (5 per child or pair); crayons for Problem B (5 per child or pair)

Problem A

José pasted 3 stars on one valentine card. He pasted some stars on a second valentine. How many stars did he paste in all?

Missing Information—Problem A

1. José pasted 4 hearts, too. (*no*)
2. José pasted 2 stars on the second valentine. (*yes*)
3. Marie pasted 4 stars on one valentine. (*no*)

Problem B

Ethan shared some crayons with his brother. He also shared some with his sister. How many crayons did he share?

Missing Information—Problem B

1. Ethan gave his brother 4 crayons. (*no*)
2. Ethan gave his sister more crayons than he gave his brother. (*no*)
3. Ethan gave his brother 2 crayons and his sister 3 crayons. (*yes*)

TEACHING ACTIONS

1. Read and discuss Problem A.
2. Ask, "What else do we need to know to solve this problem?"
3. If necessary, give the children the choices for missing information one at a time as follows: "Can we find out if we know . . . ?" Have children use counters or paper squares to supply the numbers.
4. Repeat for Problem B. Have children use crayons to supply the numbers.

27 PROCESS PROBLEM

Joan was helping pass out pink lemonade to her classmates at snack time. She carried 2 glasses of lemonade to the red table and set them in front of 2 of the children. There were 5 children at the red table. How many more glasses of lemonade did Joan need for the red table?

MATERIALS

1" counters and 1" tiles, or pattern blocks: hexagons and squares (5 of each per child); crayons

Understanding the Problem

- What was Joan doing? (*passing out glasses of lemonade*)
- How many glasses did Joan take to the red table? (2)
- Where did she put the glasses? (*in front of 2 children*)
- How many children were at the red table? (5) Do they all have glasses? (*no*)
- What are we trying to find? (*how many more glasses Joan needed to take to the red table*)

Solving the Problem

- Can you put tiles around the table to show the children? (*there should be 5*) Now draw squares for the children where you have tiles.
- Can you use counters to show the glasses of lemonade Joan put on the table? (*2 of the children should have glasses in front of them*) Now draw circles for the glasses where you have counters.
- How can you tell how many children still needed glasses? (*they don't have glasses in front of them*)
- How many more glasses did Joan need? (3)

Solution

Draw/Use a Picture

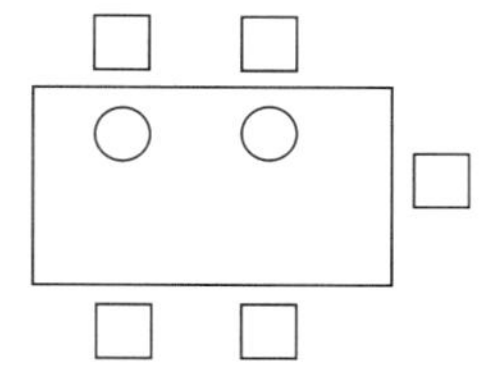

Two of the 5 children have glasses. Joan needs 3 more glasses.

Problem Extensions

1. Suppose Marco helped Joan and took a glass to the red table. How many more glasses would be needed? (2)
2. Suppose there were 6 children at the table instead of 5. If Joan passed out 2 glasses, how many more glasses would she need? (4)

STRATEGY ASSESSMENT IDEAS

Listen and watch as children work to see if they

- draw squares and circles to represent the children and glasses
- use the picture to solve the problem

28 PROCESS PROBLEM

Note: Encourage the children to solve the problem first without the aid of the picture on the blackline master. Then pass out the blackline master, and emphasize how the picture helps them solve the problem.

On Valentine's Day the children brought valentines to give to the other children in class. Mrs. Hamel, their teacher, had 2 large boxes for the valentines. One of the boxes had a heart on it. The other box had a star. Crystal put 3 valentines in the heart box and 2 in the star box. Marco put 2 valentines in the heart box and 3 in the star box. How many valentines are in each box?

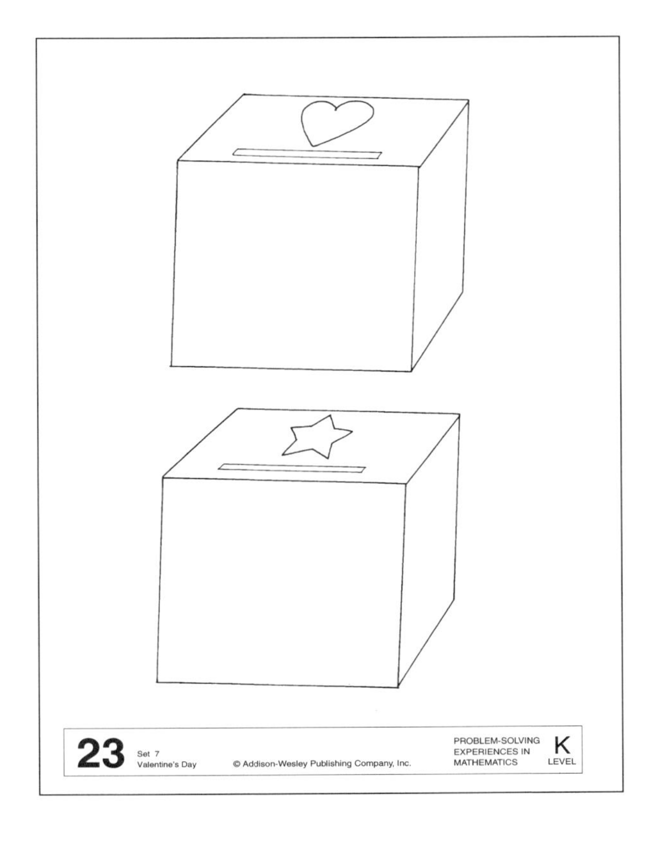

MATERIALS

1" tiles or 1" construction-paper squares (at least 10 per child); crayons

Understanding the Problem

- What did Mrs. Hamel have for the children's valentines? (*2 large boxes, 1 with a heart and 1 with a star*)
- How many valentines did Crystal put in the heart box? (3) In the star box? (2)
- How many valentines did Marco put in the heart box? (2) In the star box? (3)
- What are we trying to find? (*how many valentines are in each box*)

Solving the Problem

- How can you show the valentines Crystal put in the boxes? (*use tiles or draw them*) Have children use tiles or draw squares to show valentines in the boxes. (*3 in the heart box, 2 in the star box*)
- What else do we need to show? (*Marco's valentines*)
- How can you show the valentines Marco put in the boxes? (*use tiles or draw them*) Have children use tiles or draw squares to show valentines in the boxes. (*2 in the heart box, 3 in the star box*)
- Write the number of valentines in each box next to the box. How many valentines are in the heart box? (5) How many are in the star box? (5)

▶ *turn the page*

STRATEGY ASSESSMENT IDEAS

Listen and watch as children work to see if they

- use tiles or draw squares to represent the valentines
- use the picture to solve the problem

Solution

Draw/Use a Picture

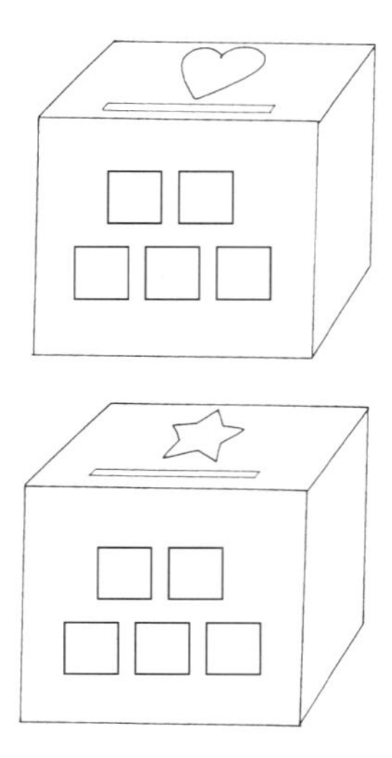

There are 5 valentines in each box.

Related Problem: 27

Problem Extensions

1. Suppose Marco put 3 valentines in both boxes after Crystal put her valentines in. How many valentines would there be in each box? (*6 in the heart box, 5 in the star box*)
2. Suppose Crystal put 1 valentine in each box, Marco put 2 in each box, and Joan put 1 in the heart box and 2 in the star box. How many valentines would be in each box? (*4 in the heart box, 5 in the star box*)

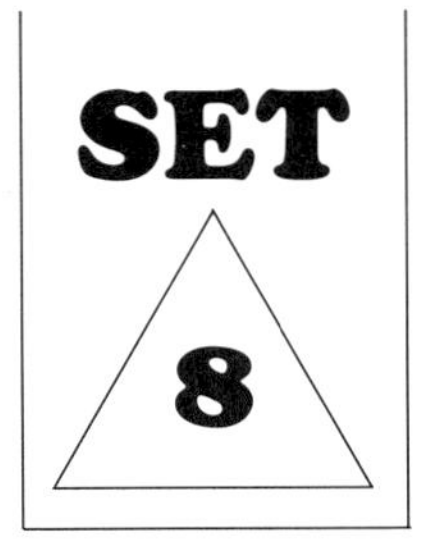

Matthew's Trip

Matthew was excited because his family was driving to see his grandma and grandpa who lived far away. His Aunt Mildred and Uncle Carll would be there, too. The best part was that Aunt Mildred and Uncle Carll would be bringing their new baby, Chip. Matthew had only seen pictures of Chip.

The car was crowded with Matthew, his brother Paul, his other brother, Joel, his mother, his father, and . . . all their luggage! Matthew and his brothers even had to bring their sleeping bags because Grandma didn't have enough beds. They would be sleeping on the floor, just like a sleep-over!

They had driven several streets away from their house when Matthew suddenly cried out, "Stop! We have to go back! We forgot Wimpy!" Wimpy was the stuffed dog he'd slept with since he was a baby.

Matthew's brothers started to tease him about needing Wimpy. Then Matthew's father reminded them about the animals *they* had had—Bonjo the monkey and Baba the bear. Matthew's father said that when he was Matthew's age, *he* had slept with a stuffed horse named Brownie. Matthew was glad his father understood!

Discussion Questions

1. Where are Matthew and his family going? (*to see his grandma and grandpa*)
2. Why was the car crowded? (*it was filled with luggage and Matthew's family*)
3. Who else was going to be at Grandma's and Grandpa's? (*Aunt Mildred, Uncle Carll, and their baby, Chip*)
4. Why did Matthew cry, "Stop!" (*he had forgotten Wimpy*)
5. Do you have brothers or sisters who tease you?
6. Do you have a stuffed animal or something else that you sleep with?

29 READINESS ACTIVITY

Act Out a Story Situation

Story A

At Grandma and Grandpa's house, all the boys built a snowperson. Matthew added 2 stones for eyes, and Paul added a carrot for his nose. Joel used 3 jar lids to make big buttons down the snowperson's tummy.

Story B

When the boys went into the house, they took off their mittens and boots and left them in the hall to dry.

TEACHING ACTIONS

1. Read and discuss Story A.
2. Have one child direct others to act out Story A.
3. Discuss what the children have acted out. Ask if it was the same as the story.
4. (*optional*) Have the children count to tell the total number of items used to decorate the snowperson. (*6*)
5. Repeat with Story B.
6. (*optional*) Have the children count to tell the total number of mittens the boys removed. (*6*)

30 CONCEPT/SKILL ACTIVITY

Complete a Picture to Show a Story

MATERIALS

1" tiles for story A (at least 5 per child or pair); 1" counters or 1" construction-paper squares for Story B (at least 3 per child or pair); crayons

Story A

Uncle Carll was a pilot. He took Matthew and his brothers to the airport to see the planes. While they were at the airport, a plane landed. Joel noticed that the plane had 1 window between the plane's propeller and its wings. There were 2 windows over the wings. The were also 2 windows behind the wings, toward the tail of the plane.

Story B

Paul said, "The plane isn't full. I can see one person in the window in front of the wings. I can see people in only 2 of the other windows."

TEACHING ACTIONS

1. Hand out Blackline Master 25. Read Story A to the children. Have them place tiles to show the windows. (1 in front of the wings, 2 over the wings, 2 behind the wings) Then have them draw small squares in place of the tiles.
2. Show several children's pictures to the class.
3. Read Story B to the children. Have them place counters to show the people. (1 in the window in front of the wings, 2 in 2 other windows) Then have them draw circles in the windows in place of the tiles.
4. Show several children's pictures. Compare different arrangements of people. Is there a person in the window in front of the wings? Are there people in 2 of the other 3 windows?
5. (*optional*) Have the children tell the total number of windows and people in the plane. (*5 windows; 3 people*)

31 PROCESS PROBLEM

The children at Pheng's table were telling about their trips and all the different ways they had traveled. They decided to keep track of all the kinds of transportation they had used. School buses wouldn't count, since they took those every day. Pheng had been on a plane, a train, and a bus. Gill had been on a plane and a bus. Jessica had been on a plane. What was the most popular way to travel? What was the least popular way?

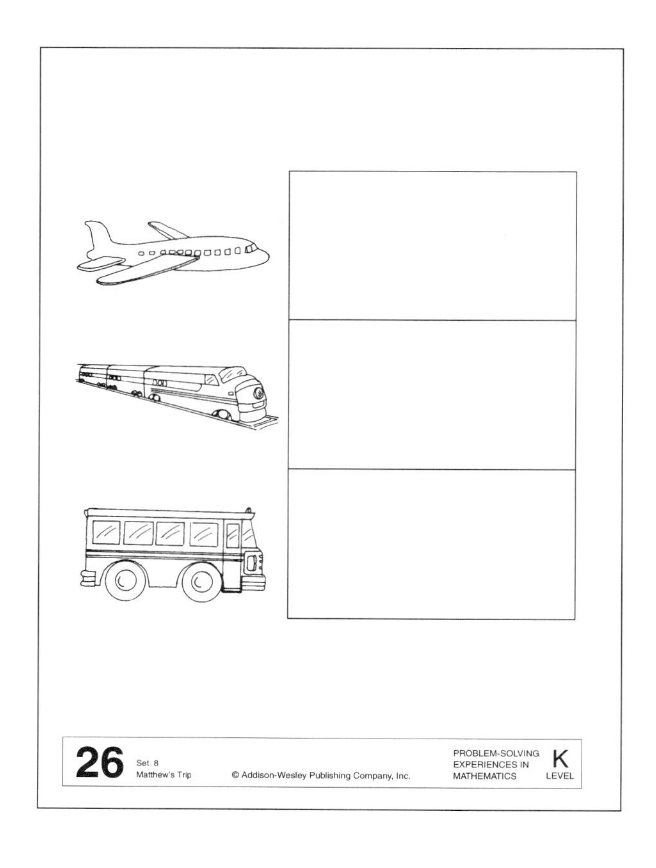

MATERIALS

interlocking cubes (at least 6 per child or pair)

Understanding the Problem

- What were the children talking about? (*ways they had traveled, kinds of transportation*)
- How had Pheng traveled? (*by plane, train, and bus*)
- How had Gill traveled? (*by plane and bus*) How had Jessica traveled? (*by plane*)
- Look at the table. What are the rows labeled? (*train, plane, bus*)
- What are we trying to find? (*what kind of transportation more children used*)

Solving the Problem

- Can you put a cube on the graph for each kind of transportation Pheng used? (*plane, train, bus*) Make sure the children line the cubes up at the left.
- Can you snap on another cube for each kind of transportation Gill used? (*plane and bus*)
- How can you show the kind of transportation Jessica used? (*snap on a cube for plane*)
- How can you tell the most popular way to travel for these children? (*it's the longest line, the plane*)
- How can you tell the least popular way to travel? (*it's the shortest line, the train*)

Solution

Complete a Graph

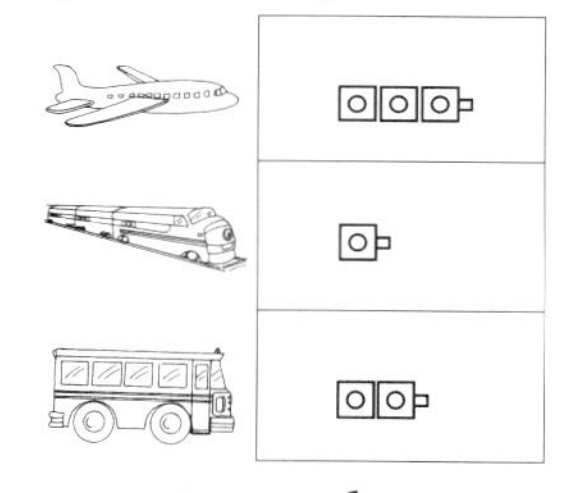

The most popular way to travel is by plane. The least popular is train.

STRATEGY ASSESSMENT IDEAS

Listen and watch as children work to see if they

- place the cubes correctly in the graph
- interpret the graph by comparing cube line lengths to arrive at the correct answers

Problem Extensions

1. Suppose Jessica had traveled by bus, too. What would the results be? (*plane: 3, train: 1, bus: 3*)
2. Have the class make predictions, then add their transportation cubes to a class table. What can they tell from the table? Were their predictions right?

32 PROCESS PROBLEM

Matthew's grandma has a weather station at her house. She charts the weather, temperature, and wind speed each day. Matthew has been keeping track of the weather at his house so he can compare the weather at his house with his grandma's.

Each day he puts another square on his weather line: yellow for sunny (have children put a yellow cube on sunny days), white for snowy, brown for cloudy, or black for rainy. (Have children place cubes on the appropriate symbols on the weather line.)

At the end of two weeks, Matthew could see there were 3 snowy days. What was the weather like most often? The next most often? What kind happened the least?

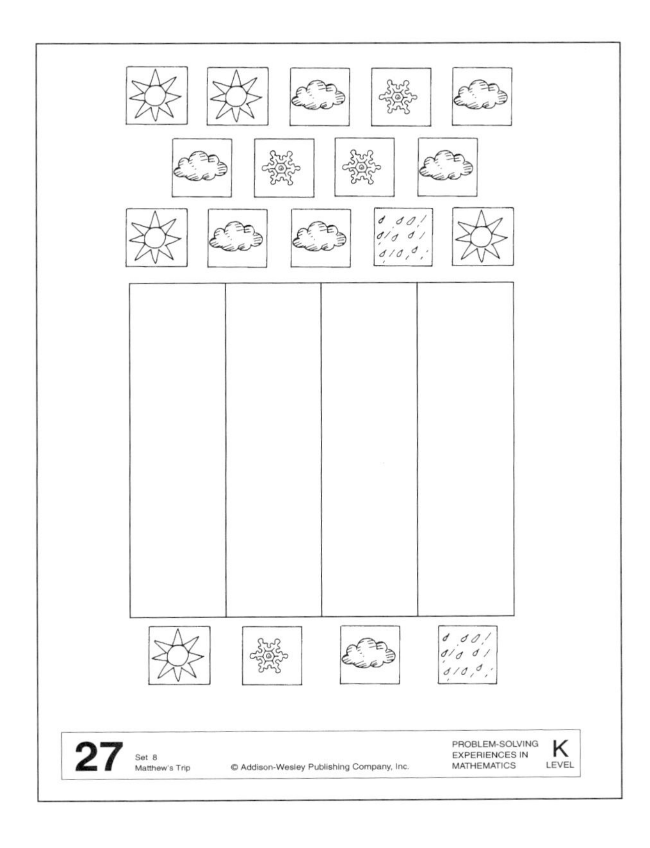

MATERIALS

interlocking cubes (at least 4 yellow, 3 white, 6 brown, and 1 black per child or pair); yellow, white, brown, and black crayons

Understanding the Problem

- What does Matthew's grandma do? (*she has a weather station and keeps track of the weather*)
- How has Matthew been keeping track of the weather? (*he put squares on a weather line*)
- What was Matthew going to do with his information? (*compare the weather at his house with his grandma's*)
- What are we trying to find? (*what was the weather like most often, the next most often, and the least*)

Solving the Problem

- Look at the graph. How are the columns labeled? (*with symbols for sunny, snowy, cloudy, rainy*)
- How could we find out how many sunny days there were? (*move the yellow cubes from the weather line to the graph*) Snowy days? (*move the white cubes to the graph*) Can you fill in the rest of the graph? Make sure children align their cubes at the bottom of the graph.

▶ *turn the page*

STRATEGY ASSESSMENT IDEAS

Listen and watch as children work to see if they

- place the cubes correctly in the graph
- interpret the graph by comparing cube line lengths to arrive at the correct answers

- How can you tell what the weather was like most often? (*it's the column with the most cubes, cloudy days*)
- What else do we need to find? (*what weather happened the next most often*) How can you tell? (*it's the column with the next most number of cubes, sunny days*)
- What was the weather like least? (*rainy*) Write the numbers of cubes under each column, then draw and color squares for each cube.

Solution

Complete a Graph

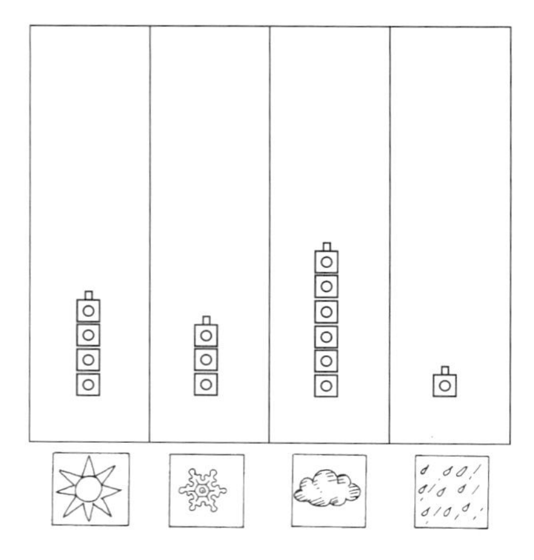

The most frequent weather was cloudy. The next was sunny. The least frequent was rainy.

Related Problem: 31

Problem Extensions

1. Suppose the next 3 days were sunny. What weather would have happened the most often? (*sunny*) The next most often? (*cloudy*)
2. Were there more wet days or dry days? (*10 dry days, 4 wet days*)

Stuffed Animal Day

Friday was going to be Stuffed Animal Day in Ms. Pope's room. All week the children had been making plans and talking about their favorite animals. On Wednesday Ms. Pope asked them to guess how many different kinds of animals would be brought. Sarah thought that since there were 25 children, there would be 25 different kinds. Hans knew he and Van were both bringing dogs. He guessed 20 kinds. Anna knew many of her friends had bears. She guessed only 10 different kinds of animals would be brought.

Chantal still didn't know what animal to bring. She had so many and she liked them all! She asked her mother, her father, and her sister what to do and they each gave her a different answer!

On Friday the children each brought their animals and introduced them during Circle Time. Then Ms. Pope said, "Let's sort them. How could we do it?" The children found many ways: by size, by color, by clothes, by furriness, and finally . . . by kind. Some of the animals were dogs, cats, monkeys, and horses. There was even an alligator! When the children counted the groups, they found that no one had guessed correctly. Can you guess how many kinds of animals were? There were 12 different kinds of animals brought for Stuffed Animal Day!

Discussion Questions

1. Why was Friday a special day? (*it was Stuffed Animal Day*)
2. What did Ms. Pope ask the children to guess? (*how many different kinds of animals would be brought*)
3. Why did Sarah think there would be 25 different kinds? (*there were 25 children*)
4. Why did Anna think there would only be 10 kinds? (*many of her friends had bears*)
5. How many different kinds of animals do you think our class would bring?

33 READINESS ACTIVITY

Retell a Number Story

Story

At Vang's table, 3 children brought bears and
1 brought a mouse.

Task 1

Retell the story with different numbers of animals.

Task 2

Retell the story with different kinds of animals.

TEACHING ACTIONS

1. Read and discuss the story.
2. Give Task 1 to the children. Model it, then solicit a variety of stories.
3. Repeat for Task 2.

34 CONCEPT/SKILL ACTIVITY

Choose Missing Data

MATERIALS

plastic links (at least 5 per child) and paper clips (at least 3 per child) for Problem A; interlocking cubes (at least 5 single cubes and 3 double—snapped together in groups of 2 per child) for Problem B

Problem A

Hans' dog had long ears. Some of the other dogs had long ears, too. How many had long ears?

Missing Information—Problem A

1. Three dogs had short ears. (*no*)
2. Van's dog had long ears. (*no*)
3. Four other dogs had long ears, too. (*yes*)

Problem B

Some of the bears were big and some were small. Were there more big bears or more small bears?

Missing Information—Problem B

1. There were 2 big bears. (*no*)
2. There were 6 small bears. (*no*)
3. There were 3 big bears and 5 small bears. (*yes*)

TEACHING ACTIONS

1. Read and discuss Problem A.
2. Ask, "What else do we need to know to solve this problem?"
3. If necessary, give the children the choices for missing information one at a time as follows: "Can we find out if we know . . . ?" Have children use plastic links for dogs with long ears and paper clips for dogs with short ears to supply the numbers.
4. Repeat for Problem B. Have children use single and double cubes to supply the numbers.

35 PROCESS PROBLEM

Chantal finally decided to bring her stuffed cat, Cuddles. Then she had to decide what Cuddles should wear. Chantal had a red ribbon and a blue ribbon she could tie around Cuddles' neck. There were two hats Cuddles could wear; one was green and one was yellow. After trying hats and ribbons on Cuddles, Chantal found 4 different ways of putting a hat and a ribbon together. Can you find the 4 different ways?

MATERIALS

red and blue plastic links or lengths of yarn; green and yellow 1" counters or 1" construction-paper squares (at least 3 of each color per child); red, blue, green, and yellow crayons

Understanding the Problem

- What can Cuddles wear? (*a ribbon and a hat*)
- How many ribbons are there? (2) What colors are they? (*red, blue*) Color the ribbons Chantal is holding.
- How many hats are there? (2) What colors are they? (*green, yellow*) Color the hats Chantal is holding.
- How many different ways did Chantal find to put a ribbon with a scarf? (4) Point to the 4 pictures of Cuddles wearing a ribbon and a hat. What are we trying to find? (*the 4 different ways to put a ribbon with a hat*)

Solving the Problem

- Suppose you start with a red ribbon. Put a red link on one of the pictures of Cuddles to show a red ribbon. What color hat can you choose? (*green or yellow*) Use a green or yellow counter to show the color hat you chose to go with the red ribbon.
- Suppose you use red ribbon again. Put a red link on another picture. Can you use the same color hat as before? (*no*) Why not? (*this would be the same as the first way*) Use a counter to show the second hat.
- If we use the red ribbon again is there a different hat we could use? (*no*) What color of ribbon do we need to use now? (*blue*) Can you finish the last 2 pictures?
- When you're sure you have 4 different ways, color the pictures to match the links and counters.

STRATEGY ASSESSMENT IDEAS

Listen and watch as children work to see if they

- name ribbon and hat combinations
- organize the search for combinations (e.g., find all combinations with the red ribbon first)
- find all possible combinations

Solution

Complete an Organized List

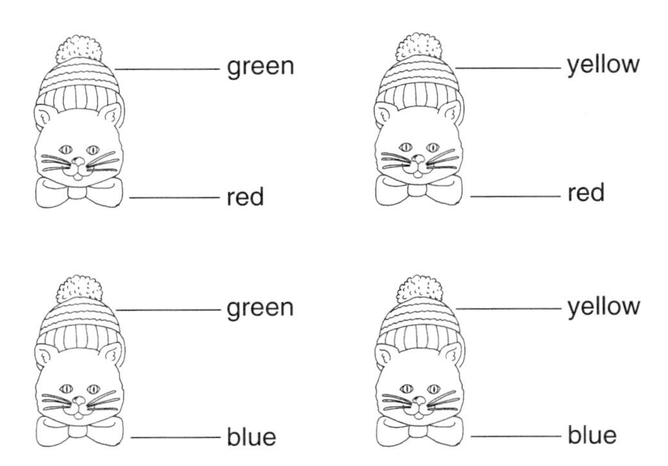

The combinations are red ribbon and green hat, red ribbon and yellow hat, blue ribbon and green hat, and blue ribbon and yellow hat.

Problem Extensions

1. Suppose the ribbons were orange and white and the hats were purple and red. How many ways could Chantal put them together? (*4*)
2. If Chantal had another hat, a white one, how many ways can she put the ribbons and hats together? (*2 more ways, or 6 in all*)

36 PROCESS PROBLEM

The children had made shape place mats for snack time. Some of the place mats had triangles on them, and some had squares. The snacks were raisins and apple slices. Chang was the helper, and was giving out snacks. When he was done, he said, "There are 4 ways to put place mats and snacks together." Can you find the 4 ways?

MATERIALS

pattern blocks: triangles and squares for the mats, trapezoids for apple slices, and black or brown 1" counters or raisins (at least 3 of each per child); green, orange, red, and brown crayons

Understanding the Problem

- How many kinds of place mats are there? (2) What shapes do they have on them? (*triangles, squares*) Draw a triangle on one of the place mats above Chang, and draw a square on the other.
- How many kinds of snacks are there? (*2; raisins and apple slices*) Color the raisins brown and the apple slices red.
- Point to the 4 rectangles and explain that these are place mats. How many different ways did Chang find to put a snack with a place mat? (4) What are we trying to find? (*the 4 different ways*)

Solving the Problem

- Suppose you started with a triangle place mat. Put a triangle on one of the place mats to show the design. What snack can you choose? (*raisin or apple*) Choose a red trapezoid for an apple slice or a counter for a raisin, and put it on the triangle place mat.
- Suppose you use a triangle place mat again. Put a triangle on one of other the mats. Can you use the same snack with it? (*no*) Why not? (*there are 4 different ways; this would be the same as the first way*) Put a different snack on the second triangle mat.

- If we use a triangle mat again, is there different snack we could use? (*no*) What kind of mat do we need to use now? (*one with a square*) Can you finish the last 2 mats?
- When you're sure you have 4 different ways, draw the designs and snacks on the mats.

STRATEGY ASSESSMENT IDEAS

Listen and watch as children work to see if they

- name place mat and snack combinations
- organize the search for combinations (e.g., find all combinations with the triangle mats first)

Solution

Complete an Organized List

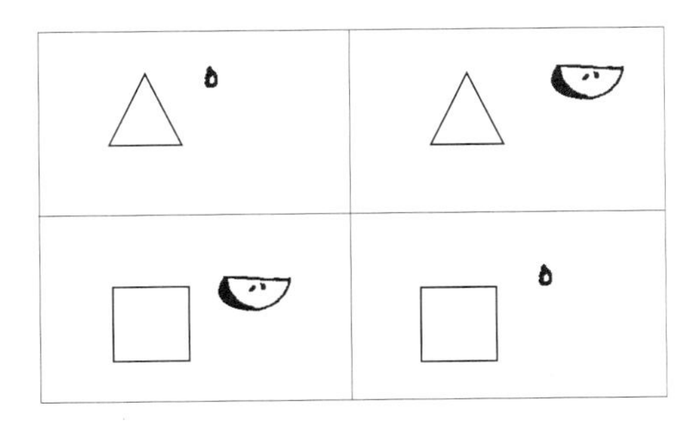

The combinations are triangles and apples, triangles and raisins, squares and apples, and squares and raisins.

Related Problem: 35

Problem Extensions

1. Suppose the mats were red and green with no designs. How many ways could Chang put the mats and snacks together? (4)
2. Suppose Chang puts a yellow or green crayon on the shape mats. How many ways can he put the crayons and mats together? (4)

Signs of Spring

Gretchen and her friend Choua were standing at the bus stop waiting for the school bus. It was such a dark and rainy morning that all the cars had their lights on. Their windshield wipers were going back and forth. All of a sudden there was a strong wind. It blew the rain right under Gretchen's rain hat and onto her face! "Oh! I didn't know I'd be swimming this morning!" Gretchen said. "This is better weather for ducks than for people."

Choua looked at the pond across the street and said, "You're right about that—look at those 2 ducks swimming in the rain. They don't have to wear raincoats the way we do!" Both girls laughed to think of the ducks wearing raincoats and rain hats.

At Sharing Time that morning, Gretchen told the class about the strong wind, the rain, and the ducks without raincoats. Lupe showed a flower from his yard that had started to bloom. Pierce said that a robin was building a nest in the tree next to his bedroom window. Mr. Bonkovsky said that the things the children had shared meant that something was coming. Could they guess what it was? Can you guess? The wind and rain, the new flower, and the robin building a nest all were signs that spring was coming!

Discussion Questions

1. What was the morning like outside? (*it was dark, rainy, and windy*)
2. What happened to Gretchen? (*the wind blew rain onto her face*)
3. Why did Gretchen and Choua laugh? (*they thought the idea of ducks wearing raincoats was funny*)
4. Where was a robin building a nest? (*in a tree next to Pierce's window at home*)
5. What did Mr. Bonkovsky say was coming? (*spring*)
6. What signs of spring can you think of?

37 READINESS ACTIVITY

Act Out a Story Situation

Story A

All the people at the city bus stop were holding umbrellas to keep dry. There were 3 women and 2 men. When a bus came along, the women closed their umbrellas and got on the bus. The men were waiting for a different bus.

Story B

When the next bus came, a woman got off, put up her umbrella, and stood with the men waiting for their bus.

TEACHING ACTIONS

1. Read and discuss Story A.
2. Have one child direct others to act out Story A.
3. Discuss what the children have acted out. Ask if it was the same as the story.
4. (*optional*) Have the children count to tell the total number of people at the bus stop before and after the bus came. (*5 before; 2 after*)
5. Repeat for Story B.
6. (*optional*) Have the children count to tell the number of people left at the bus stop. (3)

38 CONCEPT/SKILL ACTIVITY

Complete a Picture to Show a Story

MATERIALS

1" counters or tiles for story A (at least 1 orange and 3 red per child or pair); green plastic links for Story B (at least 8 per child or pair); orange, red, and green crayons

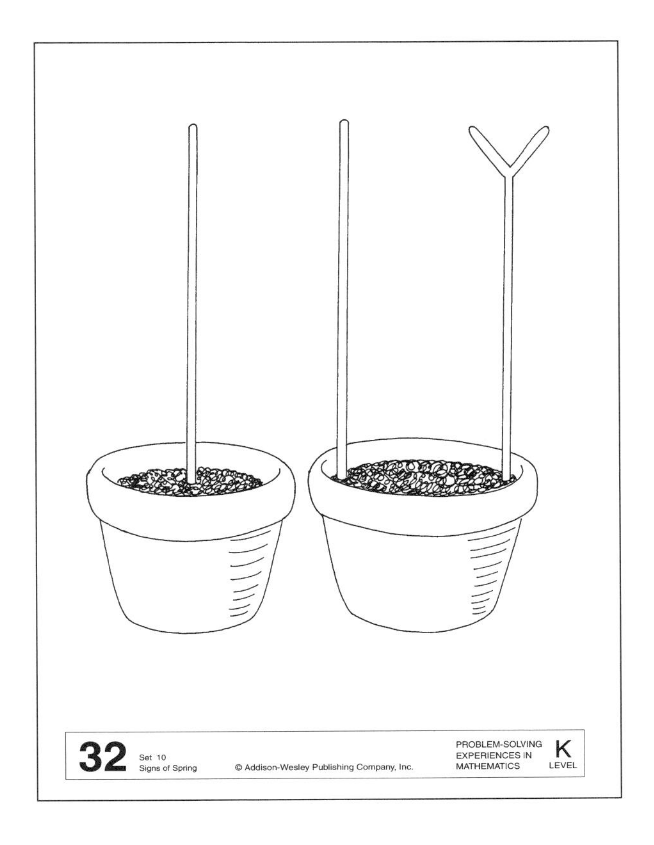

Story A

Mr. Bonkovsky had two flowerpots on the window sill. The plant in the first pot had 1 stem with an orange flower. There were 2 plants in the other pot. One plant's stem had a red flower. The other plant's stem had 2 red flowers.

Story B

The stem with an orange flower had 2 leaves growing from it. The plant with 1 red flower had 2 leaves. The plant with 2 red flowers had 4 leaves.

TEACHING ACTIONS

1. Hand out Blackline Master 32. Read Story A to the children. Have them use orange and red counters to show the flowers. Then have them draw red and orange flowers.
2. Show several children's pictures to the class.
3. Read Story B to the children. Have them place links to show the leaves. Then have them draw the leaves on the stems.
4. Show several children's pictures. Compare different arrangements of flowers and leaves. Does one pot have 1 stem, 1 orange flower, and 2 leaves? Does the other pot have 1 stem with 1 red flower and 2 leaves, and the other stem with 2 red flowers and 4 leaves? Do the pictures show the story even though the flower and leaf shapes might be different?
5. (*optional*) Have the children tell the total number of flowers and leaves in the pots. (*4 flowers; 8 leaves*)

39 PROCESS PROBLEM

After school, Jon and Choua went to look at the ducks. There were 5 ducks in all on the bank of the pond. Choua saw a group of them go into the water at one end of the pond. Jon saw the rest of them start swimming at the other end. There was 1 more duck in Choua's group than in Jon's. How many ducks were in each group?

MATERIALS

interlocking cubes (5 per child or pair); crayons

Understanding the Problem

- What were Jon and Choua doing? (*looking at ducks*)
- How many groups of ducks did they see? (2)
- How many ducks were there in all? (5) Have children count out 5 cubes and place them on the ducks on Blackline Master 33.
- What do we know about Choua's group? (*Her group had 1 more duck than Jon's group.*)
- What are we trying to find? (*how many ducks were in each group*)

Solving the Problem

- How many ducks do you think were in each group? Elicit guesses.
- Could we have 1 duck in one group and 2 in the other? Have children put 2 cubes at one end of the pond and 1 at the other end. (*no, we don't have enough ducks; there are 5 ducks in all*)
- Could we have 3 ducks and 1 duck? Have children snap another cube onto the 2 cubes at one end of the pond. (*no, there was only 1 more in Choua's group*)
- Can you guess the number in each group and use your cubes to check your guesses? Then draw the ducks in each group.

Solution

Guess and Check

There are 3 ducks in Choua's group and 2 in Jon's group.

STRATEGY ASSESSMENT IDEAS

Listen and watch as children work to see if they

- make a first guess that indicates an understanding of the problem (1 more duck in one group than in the other; a total of 5)
- make a second guess using what they learned from checking with the cubes
- use cubes to check their guesses

Related Problems: 16, 15, 4, 3

Problem Extensions

1. Suppose 1 duck swam to the other group. How many could be in each group? (*4 in Choua's and 1 in Jon's, or 2 in Choua's and 3 in Jon's*)
2. Suppose there were 3 more ducks in Choua's group than in Jon's. How many would be in each group? (*4 in Choua's; 1 in Jon's*)

40 PROCESS PROBLEM

The children in Mr. Bonkovsky's class were making diamond-shaped kites with tails. Gretchen, Choua, and Jon each made a kite. They had 9 bows in all to put on the kites' tails. They want to make sure each kite has the same number of bows. How many should they put on each kite's tail?

MATERIALS

plastic links (9 per child or pair); crayons

Understanding the Problem

- What was Mr. Bonkovsky's class doing? (*making kites*)
- Where will the children put the bows? (*on the tails*)
- How many kites are there in all? (3) How many bows? (9) Have children count out 9 links and place them on the bows on Blackline Master 34.
- What do the children want to do? (*put the same number of bows on each kite*)
- What are we trying to find? (*how many bows will go on each kite*)

Solving the Problem

- How many bows do you think will go on each kite? Elicit guesses.
- Could we have 1 bow on each kite? Have children put 1 link on each kite tail. (*no, we haven't used enough bows; there are 9 bows in all*)
- Could we add 4 more bows to one kite and 2 to another? Have children add these links. (*no, each kite gets the same number of bows*)
- Can you guess the number of bows that should go on each kite and use your links to check your guess? Then draw the bows on the kites.

Solution

Guess and Check

There are 3 bows on each kite's tail.

STRATEGY ASSESSMENT IDEAS

Listen and watch as children work to see if they

- make a first guess that indicates an understanding of the problem (an equal number of bows on each kite; a total of 9 bows)
- make a second guess using what they learned from checking with the links
- use links to check their guesses

34 Set 10 Signs of Spring © Addison-Wesley Publishing Company, Inc. PROBLEM-SOLVING EXPERIENCES IN MATHEMATICS K LEVEL

Related Problems: 39, 16, 15, 4, 3

Problem Extensions

1. Suppose there were 7 bows instead of 9. Could each kite have the same number of bows? (*no*)
2. Suppose there were 4 kites and 8 bows. How many bows would go on each kite? (2)

Raymond's Birthday

This Friday is Raymond's birthday, and he's very excited. Last week Ms. Lee sent home her "News from Kindergarten" calendar, and Raymond's name was in one of the squares: "Raymond's Birthday." Raymond circled the square with a red crayon for everyone in his family to see. His mother used a magnet to hang the calendar on the refrigerator.

In school Ms. Lee will have the whole class sing "Happy Birthday" to Raymond during Circle Time. Next she'll put the birthday crown on his head and a sticker on his shirt that says "Happy Birthday Raymond." Raymond hopes it will be a sunny day so they can go outside for recess. Then his best friend Rochanne in the other kindergarten class can see him wearing the crown. During snack time Raymond will pass around special treats for the children. He hasn't decided yet what he will bring. His favorite cookies are oatmeal-raisin, but he also likes fruit rolls, and they come in different flavors.

On Friday at breakfast his family will give him his presents. His big sister, Danielle, has been teasing him for a week about her gift. One day she told him she's going to give him a bag of pebbles, and just yesterday she said her present would be a small plant. Raymond doesn't think he wants either of these, but he knows Danielle always gives him something special. She seems to know just what he will like. Raymond is sure this will be his best birthday ever!

Discussion Questions

1. Why is Raymond excited? (*this Friday is his birthday*)
2. What will Ms. Lee do for Raymond during Circle Time? (*have the class sing "Happy Birthday"; put a crown on his head*)
3. Why does Raymond hope it will be sunny for his birthday? (*so the class can go out for recess*)
4. What do you think Danielle will give Raymond?
5. How do you celebrate your birthday?

41 CONCEPT/SKILL ACTIVITY

Tell a Question Using Data in the Story

MATERIALS

plastic links, pencils, or crayons (at least 3 red and 2 blue per child or pair)

Story A

Ms. Lee always lets the birthday boys and girls choose a shiny pencil from her pencil box. Today she took out 3 red pencils and then 2 blue pencils.

Possible Questions for Story A

How many pencils has Ms. Lee taken out?

Story B

Raymond is passing out fruit rolls. He has 2 strawberry rolls and 1 blueberry roll left.

Possible Questions for Story B

How many fruit rolls does Raymond have left?

TEACHING ACTIONS

1. Read and discuss Story A.
2. Read the story again, and have children use pencils or crayons to show the pencils.
3. Ask children to tell a question that can be answered using the numbers in the story.
4. Repeat for Story B.
5. (*optional*) Have children tell how many pencils Ms. Lee has taken out in Story A (*5*) and how many fruit rolls Raymond has left in Story B (*3*).

42 CONCEPT/SKILL ACTIVITY

Complete a Picture to Show a Story

Picture

MATERIALS

drawing paper; pattern blocks: squares (at least 8 per child or pair for Story A); pattern blocks: triangles (at least 5 per child or pair); blue 1" counters or 1" construction-paper squares (at least 3 per child or pair) for Story B; orange, blue, and green crayons

Story A

Raymond and Danielle are getting the table ready for his party. They will need 4 chairs on one side of the table and 4 on the other side.

Story B

Each guest will get a hat to wear. There are 5 hats with pointed tops and 3 hats with round tops.

TEACHING ACTIONS

1. On the board, draw the rectangle shown. Have the children copy the picture. Tell them that the rectangle is a table.
2. Read Story A to the children. Have them place square pattern blocks outside the rectangle to show chairs. Then have them draw small squares to replace the blocks.
3. Sketch several children's pictures on the board.
4. Read Story B to the children. Have them place triangles for pointed hats and counters for round hats on the table. Then have them draw these to replace the counters.
5. Sketch several children's pictures on the board. Compare different arrangements of hats. Are there 5 triangles and 3 circles in each picture even though they are in different places?
6. (*optional*) Have the children tell the total number of hats on the table. (*8*)

43 PROCESS PROBLEM

The day of the party has finally come. Raymond has red, blue, and green balloons. He will tie one balloon for each guest to the stair railing in front of his building. Raymond decides to make a pattern with the colors. He ties on a red balloon first, a blue balloon next, and a green balloon third. Next comes a red balloon, then a blue balloon. Can you decide what balloons Raymond will choose next? (Have children place counters on the balloons in the picture—red, blue, green, red, blue.)

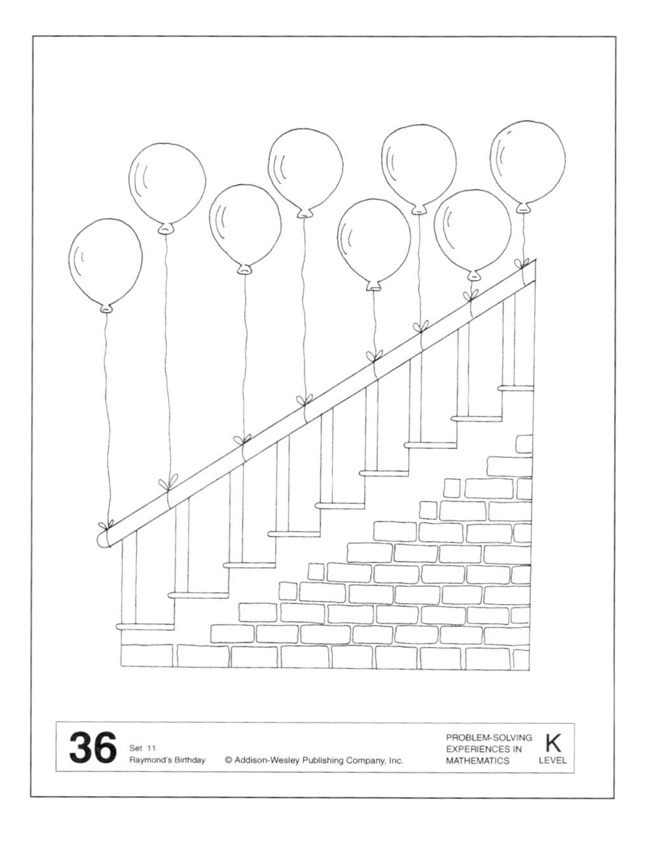

MATERIALS

red, blue, and green 1" counters (at least 3 of each per child or pair); red, blue, and green crayons

Understanding the Problem

- What is Raymond tying on the railing? (*balloons*)
- What colors are his balloons? (*red, blue, and green*)
- What are we trying to find? (*which balloon comes next*)

Solving the Problem

- How is this balloon (point to the second balloon) different from the first balloon (point to the first balloon)? (*blue, not red*)
- Repeat the first question for balloons 3 and 2, 4 and 3, 5 and 4.
- What color will the next balloon be? (*green*) How do you know? (*the balloon colors form a pattern*)
- Have children place the green counter and remove the counters to color the balloons.
- What color will the next balloons be? (*red, blue*) Have children color the last two balloons.

Solution

Look for a Pattern

Pattern: red, blue, green, red, blue, green, red, blue

Raymond will choose a green, red, then blue balloon.

Related Problems: 20, 19, 8, 7

Problem Extensions

1. Suppose the colors were blue, green, red, blue, green. What color of balloon will be next? (*red*)
2. Have children place counters in the following pattern: blue, green, red, blue, (space), (space), blue, green, red. Explain that the wind blew two of the balloons away. What colors were they? (*green, red*) Have children place counters for the missing balloons.

STRATEGY ASSESSMENT IDEAS

Listen and watch as children work to see if they

- describe that the balloon colors form a pattern
- extend the pattern with the correct colors of counters
- use the pattern to arrive at a correct solution to the problem

44 PROCESS PROBLEM

Raymond could hardly wait to show his friends what his big sister, Danielle, gave him for his birthday. She did give him a bag of pebbles and a small plant. The big surprise was that they were for the bottom of a goldfish bowl! The bowl even had a goldfish in it. Raymond is going to let his friends name the goldfish.

The party is on Saturday so Raymond's father and Aunt Mayla can be there. Aunt Mayla teaches math at the high school during the week, and she's also a magician. She shows the boys and girls an empty box, then she reaches in and s-l-o-w-l-y pulls out . . . a yellow scarf! Raymond's friends are surprised and say, "Oooh!" Aunt Mayla keeps pulling and tied to the first scarf is . . . an orange scarf! Tied to that scarf is a blue one, then a red one. After another yellow scarf and an orange one, Aunt Mayla tells Raymond and his friends that there are two scarves left. She says they ought to be able to guess the colors. Do you know what colors they are? (Have children fit the links together in order—yellow, orange, blue, red, yellow, orange.)

MATERIALS

yellow, orange, blue, and red plastic links (at least 3 each of per child or pair); yellow, orange, blue, and red crayons

Understanding the Problem

- What colors are the scarves Aunt Mayla is pulling from the box? (*yellow, orange, blue, and red*)
- How many scarves are there in all? (*8*)
- What are we trying to find? (*what color scarves will come next*)

Solving the Problem

- How is this second scarf (point to the second scarf) alike or different from the first scarf (point to the first scarf)? (*the first is yellow, the second is orange*)
- Repeat the first question for scarves 3 and 2, 4 and 3, 5 and 4, 6 and 5.
- What color will the next scarf be? (*blue*) How do you know? (the scarves form a color pattern)
- Can you add a link for the last scarf? (*red*)

▶ *turn the page*

STRATEGY ASSESSMENT IDEAS

Listen and watch as children work to see if they

- describe that the scarf colors form a pattern
- extend the pattern correctly with links
- use the pattern to arrive at a correct solution to the problem

Solution

Look for a Pattern

Pattern: yellow, orange, blue, red, yellow, orange, blue, red

The next scarf is blue. The last scarf is red. Have children color the scarves to match the links.

Related Problems: 43, 20, 19, 8, 7

Problem Extension

Have children fit links together in the following pattern: blue, red, blue, red, yellow, yellow, blue, red. This time there are 12 scarves in the box and 6 in the color pattern. What colors are the last 4 scarves? (*blue, red, yellow, yellow*)

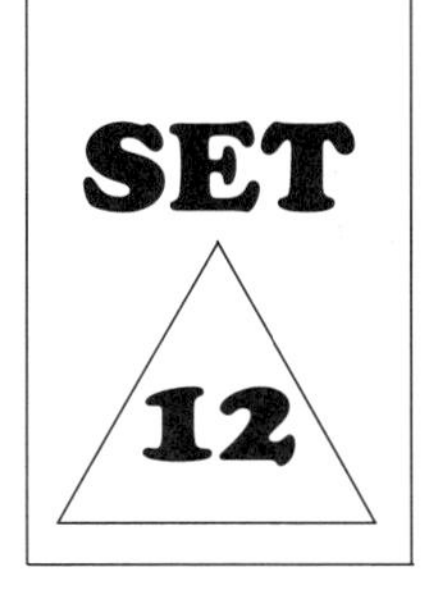

The School Fair

Scenic Heights Elementary School was going to have a fair. There were going to be games that people could play to win prizes. Each class was going to make one game. Mr. Melcher wouldn't tell his class what their game was going to be—they were going to have to guess! He did tell them that they would make the game that day.

Mr. Melcher showed the children a basket of paper punches for punching holes. He also showed them a box of paper clips. Next he held up sheets of stiff paper that had fish drawn on them. The fish had numbers on them—either 1, 2, 3, 4, or 5. First the children colored and cut out the fish. Next Mr. Melcher took one of the fish and showed the children how to punch a hole in it. Then he took a paper clip and attached it to the fish through the hole.

"What do you think our game is?" he asked. Marlis guessed that people would have to add the numbers on the fish. Then Mr. Melcher brought out a stick that had a string tied on one end. There was a magnet tied on the other end of the string. "That looks like a fishing pole!" exclaimed Francesca. Mr. Melcher gave her the pole, and she held it over her fish. The fish's paper clip stuck to the magnet on the end of the string. The fish lifted off the table! Francesca's fish had a number 4 on it. During the fair that would mean she would get prize number 4. The children could hardly wait for fair night. They were sure their game would be the most fun.

Discussion Questions

1. What was Scenic Heights Elementary school going to have? (*a fair*)

2. What was each class going to do? (*make a game for the fair*)

3. What did Mr. Melcher show the children? (*paper punches, paper clips, fish*)

4. How did the children fix the fish? (*they colored and cut them out, punched a hole in them, and put a paper clip through the hole*)

5. Why did the fish get picked up? (*the magnet on the end of the string attracted the paper clip*)

6. Have you ever played a game at a fair? What was it like?

45 CONCEPT/SKILL ACTIVITY

Choose Missing Data

MATERIALS

blue 1" counters or 1" construction-paper squares (at least 6 per child) for Problem A; 1" counters (at least 7 per child or pair) and plastic links (at least 9 per child or pair) for Problem B

Problem A

Francesca's fish was colored blue. Some of the other fish were blue, too. How many fish in all were blue?

Missing Information—Problem A

1. Three fish were yellow. (*no*)
2. Franz's fish was blue. (*no*)
3. Five other fish were colored blue. (*yes*)

Problem B

Some of the fish had big tails and some had small tails. Were there more fish with big tails or more with small tails?

Missing Information—Problem B

1. There were 9 fish with big tails. (*no*)
2. There were 5 fish with small tails. (*no*)
3. There were 7 fish with big tails and 6 fish with small tails. (*yes*)

TEACHING ACTIONS

1. Read and discuss Problem A.
2. Ask, "What else do we need to know to solve this problem?"
3. If necessary, give the children the choices for missing information one at a time as follows: "Can we find out if we know . . . ?" Have children use counters to supply the numbers.
4. Repeat for Problem B. Have children use links for big tails and counters for small tails to supply the numbers.

46 CONCEPT/SKILL ACTIVITY

Tell a Question Using Data in the Story

MATERIALS

red and blue 1" counters or construction-paper squares (at least 3 of each per child or pair)

Story A

Mr. Melcher let the children practice with the fishing pole. Allison caught 3 red fish and then 2 blue fish.

Possible Question for Story A

How many fish did Allison catch?

Story B

There were 6 fish with small tails left on the table. Kate caught 3 of them. Then she caught 1 more.

Possible Question for Story B

How many fish were left on the table?

TEACHING ACTIONS

1. Read and discuss Story A.
2. Read the story again, and have children use counters to show the fish.
3. Ask children to tell a question that can be answered using the numbers in the story.
4. Repeat for Story B.
5. (*optional*) Have children tell how many fish Allison caught in Story A (5) and how many fish were left on the table in Story B (2).

47 PROCESS PROBLEM

The night of the fair finally arrived. Allison, Marlis, and Franz all won a prize. They each were very happy with their prizes. The prizes were a plastic snake, a yo-yo, and a balloon. Franz won the yo-yo. Allison had never really liked balloons. What kind of prize did each person win?

MATERIALS

1" counters or 1" construction-paper squares (2 of each of 3 colors per child or pair); crayons

Understanding the Problem

- What did Allison, Marlis, and Franz do at the fair? (*won prizes*)
- Were they happy with their prizes? (*yes*)
- What did Franz win? (*the yo-yo*)
- Who doesn't like balloons? (*Allison*)
- What are we trying to find? (*what person won each prize*)

Solving the Problem

- Can you put the same color counter on Franz and his prize? (*yo-yo*)
- If Allison doesn't like balloons, and she liked her prize, what prize did she get? (*snake*) Choose a different color and put matching counters on Allison and her prize.
- If Franz won the yo-yo, and Allison won the snake, what did Marlis win? (*the balloon*) How can you show this? (*choose a different color and put matching counters on Marlis and the balloon*)
- Have the children draw lines connecting each person to their prize.

Solution

Use Logical Reasoning

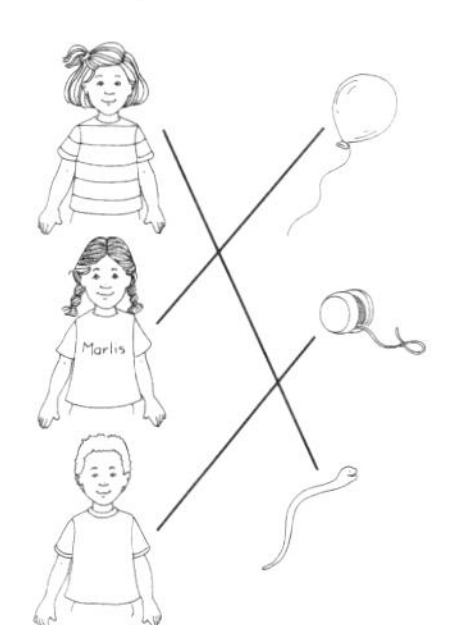

Related Problems: 24, 23, 21, 11

Problem Extension

If Marlis had gotten the yo-yo instead of the balloon, what prize would Franz have gotten? (*the balloon*)

STRATEGY ASSESSMENT IDEAS

Listen and watch as children work to see if they

- use colored counters to match people with their prizes
- correctly use all the statements in the problem
- conclude that the remaining person and the remaining prize go together

48 PROCESS PROBLEM

Some of the fourth graders dressed up as clowns for the fair. Kate's brother Ben was one of them, but Kate's friends didn't recognize him. Kate said she'd give her friends some clues. "He doesn't have a big nose. He doesn't have triangles around his eyes. He isn't a sad-face clown," she said. Can you guess which clown is Ben?

MATERIALS

1" counters or 1" construction-paper squares (4 per child or pair); crayons

Understanding the Problem

- What did some of the fourth graders do? (*dressed as clowns*)
- Who was one of the clowns? (*Kate's brother Ben*)
- What was Kate going to do? (*give her friends clues so they would know which clown was Ben*)
- What are we trying to find? (*which clown is Ben*)

Solving the Problem

- Help the children identify the features of each clown, e.g., smiling or sad-faced, large or small nose. (see Blackline Master 40)
- Read the first clue to the children. Which clowns have big noses? (see picture) Put a counter on the clowns with big noses to show they are "out."
- Read the second clue. Can you show that the clown with triangle eyes is out?
- Read the third clue. Can you show that the clown with the sad face is out?
- Who is left? (*the clown with the curly hair; Ben*) Draw a ring around the clown that is Ben.

Solution

Use Logical Reasoning

Clue 1 tells us Ben is not clown 1 or clown 5.

Clue 2 tells us Ben is not clown 3.

Clue 3 tells us Ben is not clown 2 and must be clown 4.

Related Problems: 47, 24, 23, 21, 11

▶ *turn the page*

STRATEGY ASSESSMENT IDEAS

Listen and watch as children work to see if they

- use counters to eliminate clowns
- correctly use all clues given in the problem
- conclude that the remaining clown must be Ben

Problem Extensions

1. Give a different set of clues:

  a. I have a smile. (*eliminates clown 2 and clown 5*)

  b. I have triangle eyes. (*eliminates clown 4*)

  c. I have a big nose. *(eliminates clown 3; Ben is clown 1*)

2. Secretly pick one of the clowns. Have children ask yes-or-no questions. For example: Does he have a big nose? Have children use counters to eliminate clowns.

Baby Animals

It was spring, and for several weeks the children in Ms. Apple's class had been learning about baby animals. They had seen pictures and videos of sheep and their lambs and pigs and their piglets. Vaneeta even brought her cat and new kittens to school in a carrier. The kittens made tiny mewing noises.

Today the class was visiting a petting zoo. There were many pens with animals in them. The exciting thing was that the children could go right into the pens to pet the animals! Two of the pens had sheep in them. Ms. Apple explained that a mother sheep is called a ewe. The children could see that each ewe had a baby lamb. The lambs followed their mothers wherever they went.

The children didn't go into the pigpen because it was very muddy. They got to watch the pigs, though. All the children were surprised at how large the mother pig was. She was surrounded by 6 piglets.

One piglet jumped right on top of another piglet to get close to the mother. Travis was worried about the piglet on the bottom, but the little pig just squirmed out from underneath. Then he trotted over and lay down in a mud puddle! Sang said she thought that the baby needed to go into a bathtub. That made all the children laugh.

Discussion Questions

1. About what has Ms. Apple's class been learning? (*baby animals*)
2. What did Vaneeta bring to school? (*her cat and its new kittens*)
3. How did the lambs behave? (*they followed their mothers*)
4. What about the mother pig surprised the children? (*she was bigger than they expected*)
5. Why did Sang think the piglet needed to be in a bathtub? (*the piglet lay down in a mud puddle and got all muddy*)
6. Have you ever been to a petting zoo? What animals did you see there?

49 CONCEPT/SKILL ACTIVITY

Tell a Question Using Data in the Story

MATERIALS

pattern blocks: trapezoid or hexagon (1 per child or pair); 1" counters or 1" tiles (at least 7 per child or pair)

Story A

The mother pig had 4 piglets on one side of her and 2 on the other.

Possible Questions for Story A

How many piglets were near the mother pig?

Story B

The mother duck had 4 ducklings following closely behind her. There were 3 ducklings hurrying to catch up.

Possible Questions for Story B

How many ducklings were there in all?

TEACHING ACTIONS

1. Read and discuss Story A.
2. Read the story again, and have children use pattern blocks to show the mother pig and counters to show the piglets.
3. Ask children to tell a question that can be answered using the numbers in the story.
4. Repeat for Story B.
5. (*optional*) Have children tell how many piglets were with the mother pig in Story A (6) and how many ducklings there were in all in Story B (7).

50 CONCEPT/SKILL ACTIVITY

Choose Missing Data

MATERIALS

1" counters or interlocking cubes (at least 4 brown per child or pair for Problem A and Problem B, and at least 4 yellow per child or pair for Problem B)

Problem A

One of the piglets was rolling in the mud. Some of the other piglets were rolling in the mud, too. How many piglets were rolling in the mud?

Missing Information—Problem A

1. Two piglets were rolling in the mud. (*no*)
2. The smallest piglet was rolling in the mud. (*no*)
3. Three other piglets were rolling in the mud too. (*yes*)

Problem B

Some of the ducklings were quacking and some were quiet. Were there more quacking ducklings or quiet ones?

Missing Information—Problem B

1. There were 2 quacking ducklings. (*no*)
2. There were 4 quiet ducklings. (*no*)
3. There were 3 quacking ducklings and 4 quiet ones. (*yes*)

TEACHING ACTIONS

1. Read and discuss Problem A.
2. Ask, "What else do we need to know to solve this problem?"
3. If necessary, give the children the choices for missing information one at a time as follows: "Can we find out if we know . . . ?" Have children use brown counters for piglets to supply the numbers.
4. Repeat for Problem B. Have children use brown counters for quacking ducklings and yellow counters for quiet ones to supply the numbers.

51 PROCESS PROBLEM

Note: Encourage the children to solve the problem first without the aid of the picture on the blackline master. Then pass out the blackline master and emphasize how the picture helps to solve the problem.

There were 2 black sheep in the pen shaped like a circle. There were 3 black sheep in the pen shaped like a square. There were 4 gray sheep in the circle pen, and 2 gray sheep in the square pen. How many more sheep were in the circle pen than in the square pen?

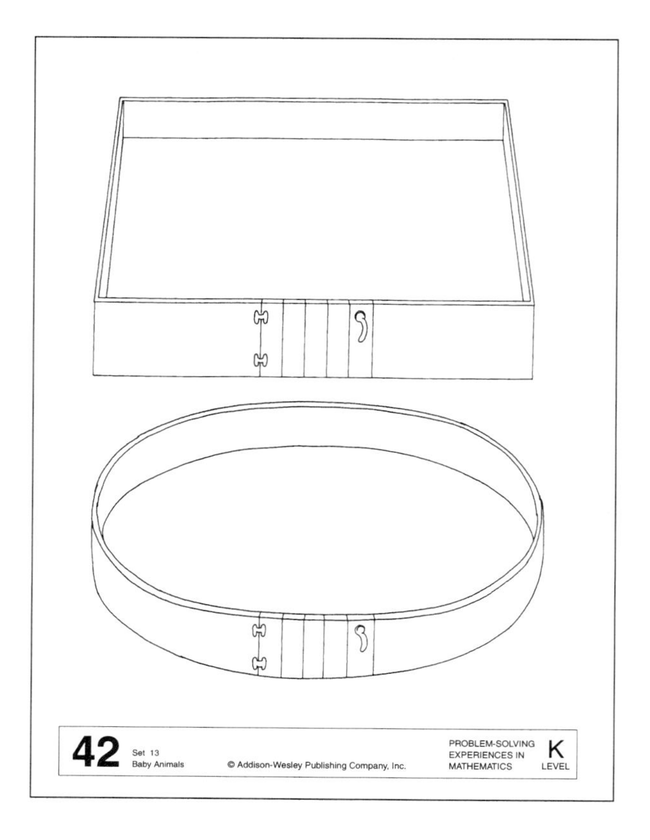

MATERIALS

1" counters, interlocking cubes, or 1" construction-paper squares (at least 5 black and 6 gray or white per child or pair); black and gray crayons

Understanding the Problem

- What color were the sheep? (*black, gray*)
- Were the pens the same shape? (*no, one was shaped like a circle; the other a square*)
- Were all the sheep in each pen the same color? (*no, there were black and gray sheep in each pen*)
- How many black sheep were in the circle pen? (*2*) The square pen? (*3*)
- What are we trying to find? (*how many more sheep are in the circle pen*)

Solving the Problem

- Can you put black counters in the pens to show the black sheep? (*2 in the circle, 3 in the square*) Now draw black circles for the black sheep where you have counters.
- How many gray sheep were in the circle pen? (*4*) The square pen? (*2*) How can you show that? (*put that many gray counters in each pen*) Now draw gray circles where you have counters.
- What are we trying to find? (*how many more sheep were in the circle pen*)
- What do we need to do? (*count the sheep in each pen*) Write the number of sheep on each pen. How many more were in the circle pen? (*1*)

STRATEGY ASSESSMENT IDEAS

Listen and watch as children work to see if they

- use colored counters and draw colored circles to represent the sheep in the 2 pens
- use the picture appropriately to solve the problem

Solution

Draw/Use a Picture

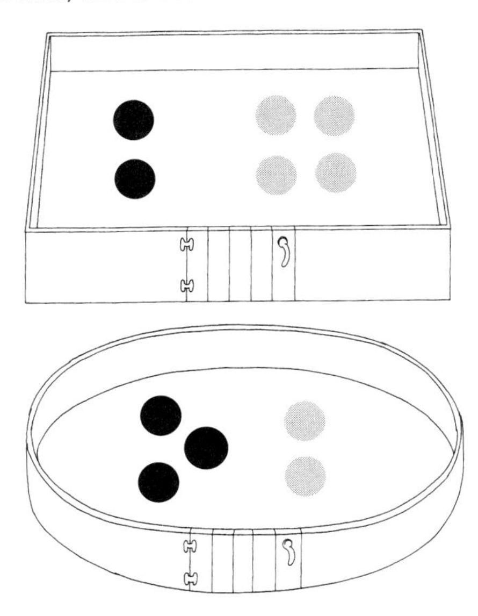

There is 1 more sheep in the circle pen than in the square pen.

Related Problems: 28, 27

Problem Extensions

1. Suppose 1 more black sheep was put into the circle pen. How many more sheep would be in the circle pen than in the square pen? (2)
2. Suppose the sheep were all gray instead of some black and some gray. How many more sheep would be in the circle pen than in the square pen? (*1*)

52 PROCESS PROBLEM

Note: Encourage the children to solve the problem first without the aid of the picture on the blackline master. Then pass out the blackline master and emphasize how the picture helps to solve the problem.

Ms. Apple had given the children bags of corn to feed the hens. Travis was feeding 3 hens. He fed the first hen 2 pieces, and the second hen 2 pieces. Then he fed the third hen 1 piece. Vaneeta fed the hens, too. The first hen ate 1 piece, the second hen ate 2 pieces, and the third hen ate 2 pieces. Which hen ate the most corn?

MATERIALS

yellow 1" counters, 1" tiles, or 1" construction-paper squares (at least 10 per child or pair); crayons

Understanding the Problem

- What did Ms. Apple give the children? (*bags of corn to feed the hens*)
- How many hens was Travis feeding? (*3*)
- Were they all eating the same number of corn pieces? (*no, each hen ate a different amount*)
- What are we trying to find? (*which hen ate the most corn*)

Solving the Problem

- How many pieces of corn did Travis feed the first hen? (*2*) How could you show that with counters? (*put 2 counters near the first hen*)
- How many pieces did he feed the second hen? (*2*) The third hen? (*1*) Show this with counters by each hen.
- How many pieces did Vaneeta feed the hens? (*first hen—1 piece; second hen—2 pieces; third hen—2 pieces*) Have children use counters to show this. Then have them draw circles to show all the pieces of corn.
- What do we need to find? (*which hen ate the most corn*) How can we tell? (*count the pieces for each hen*) Write the number of pieces of corn that each hen ate below each hen. (*3, 4, 3*) Who ate the most? (*the second hen*)

STRATEGY ASSESSMENT IDEAS

Listen and watch as children work to see if they

- use counters and draw circles to represent the pieces of corn eaten by each hen
- use the picture to solve the problem

Solution

Draw/Use a Picture

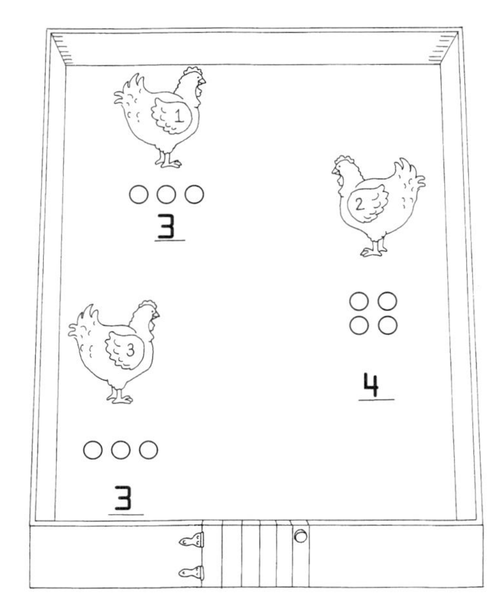

The second hen ate the most corn—4 pieces.

Related Problems: 51, 28, 27

Problem Extensions

1. What if Travis fed each hen 1 more piece. Which hen would have eaten the most? (*the second hen*)
2. Suppose the first hen ate another piece. Which hen would have eaten the most? (*the first and second hens would have eaten equal amounts—4 pieces*)

Growing Things

Mastoura woke up early and hurried to the window. Today was Saturday. Her father had said that if it didn't rain, they would plant the garden. She opened the curtain and looked out—it was a bright, sunny day! It was just right for planting.

Mastoura had always helped with her family's garden, but this year was different. Her father was going to help her plant her very own garden. She had looked at many different seed catalogs before deciding what to plant. Finally, she had chosen corn, tomatoes, carrots, and beans.

Mastoura's father helped her use the shovel to dig her garden. He said the plants would need loose soil. Next, they stretched string between sticks at each end of the garden. Below the string they made a line in the soil. The string helped them make a straight row. Finally, they put the seeds into the ground, and Mastoura covered them with soil. Some kinds of seeds should be planted deeper than others. Mastoura had to be very careful to get just the right amount of soil on top of the seeds.

At the end of the day, Mastoura was very dirty. When her father watered their gardens with the hose, he watered her, too! They both laughed and wondered if the water would help her grow as quickly as her garden.

Discussion Questions

1. Why was Saturday a special day? (*Mastoura and her father were going to plant the garden*)
2. Why was this year different? (*Mastoura would have her own garden*)
3. What did Mastoura decide to plant? (*corn, tomatoes, carrots, and beans*)
4. Why did they stretch string between 2 sticks? (*so they could plant the seeds in a straight line*)
5. Have you ever planted anything? If you could plant a garden, what would you choose to grow?

53 CONCEPT/SKILL ACTIVITY

Given a Picture, Choose an Addition Question

MATERIALS

1" counters, 1" tiles, or 1" construction-paper squares (at least 5 per child or pair)

Picture A

4 flowers in the box, 1 flower in the pot—see Blackline Master 45.

Questions for Picture A

1. How many flowers are in the box?
*2. How many flowers are there in all?

Picture B

2 flowers in the small pot, 3 flowers in the big pot—see Blackline Master 45.

Possible Question for Picture B

*1. How many flowers have been planted altogether?
2. How many flowers are in the big pot?

TEACHING ACTIONS

1. Show and discuss Picture A. Have the children use counters to represent the flowers.
2. Read the questions for Picture A. Have the children use the counters to help them choose the addition question.
3. Repeat for Picture B.
4. (*optional*) Have children tell the answer to each addition question.

54 CONCEPT/SKILL ACTIVITY

Tell the Operation

MATERIALS

1" counters, 1" tiles, or 1" construction-paper squares (at least 4 per child or pair)

Problem A

Mr. Kane was getting ready to plant bushes. He dug 3 holes in the backyard. He dug 1 hole in the front yard. How many holes did he dig altogether?

Problem B

Mr. Kane planted 3 bushes in the holes in the backyard. Then he changed his mind, and dug up 1 of the bushes. How many bushes were left in the backyard?

TEACHING ACTIONS

1. Read and discuss Problem A.
2. Have the children use counters to represent the holes Mr. Kane dug.
3. Ask children which operation is needed to solve Problem A: *putting together* or *taking apart*. Discuss how the action in the story tells which operation is needed.
4. Repeat for Problem B, having children use counters to represent the bushes.
5. (*optional*) Have the children solve each problem.

55 PROCESS PROBLEM

Mastoura's mother bought a whole box of flowers to plant in front of their house. She plans to plant 2 red striped flowers and 2 white ruffled ones in a big pot. (Have the children place red and white interlocking cubes on the striped and ruffled flowers in the box.) Then she will plant 5 blue flowers in a basket to hang from the porch. (Have the children continue to cover the flowers in the box with the appropriate cubes.) Finally, she will plant 2 red striped flowers and 1 white one in another basket. What color flower will she plant the most of?

MATERIALS

interlocking cubes (at least 4 red, 3 white, and 5 blue per child or pair); red, white, and blue crayons

Understanding the Problem

- What is Mastoura's mother doing? (*planting flowers*)
- Where is she planting the flowers? (*in a pot and in 2 baskets*)
- What flowers did she plant in the pot? (*2 red striped, 2 white*)
- What flowers did she plant in the first basket? (*5 blue*)
- Look at the table. How are the rows labeled? (*with flowers: red striped, white ruffled, and blue*)
- What are we trying to find? (*what color flower she planted the most of*)

Solving the Problem

- Can you put a cube on the graph for every flower that will go in the pot? (*2 red, 2 white*) Make sure the children align the cubes at the left.
- Can you put a cube on the graph for every flower that will go in the first basket? (*5 blue*)
- How can you show the flowers that will go in the second basket? (*snap on 2 red and 1 white cube*)
- How can you tell the color flower Mastoura's mother will plant the most of this year? (*it's the longest line, blue*)

46 Set 14 Growing Things © Addison-Wesley Publishing Company, Inc. PROBLEM-SOLVING EXPERIENCES IN MATHEMATICS K LEVEL

Solution

Complete a Graph

Mastoura's mother will plant more blue flowers this year.

Related Problems: 32, 31

Problem Extensions

1. Suppose Mastoura's mother planted 2 more red striped flowers. What color would she plant the most of? (*red*)
2. Have the class make predictions, then add cubes to a class table for their favorite flower color. What can they tell from the table? Were their predictions right?

STRATEGY ASSESSMENT IDEAS

Listen and watch as children work to see if they

- place the cubes correctly in the graph
- interpret the graph by comparing line lengths to arrive at the correct answer

56 PROCESS PROBLEM

Mastoura and her father had a list to remind them what they had decided to plant. Her father planted 4 rows of corn and 2 rows of tomatoes. (Have the children place cubes on the list: 4 yellow cubes on the corn pictures and 2 red cubes on the tomato pictures.) He also planted 4 rows of carrots. (Have the children continue to cover the pictures with cubes.) Mastoura planted 2 rows of corn and 1 row of tomatoes. Then she planted 1 row of carrots. When they were done, Mastoura asked, "What vegetable is planted in the most rows? The fewest rows?"

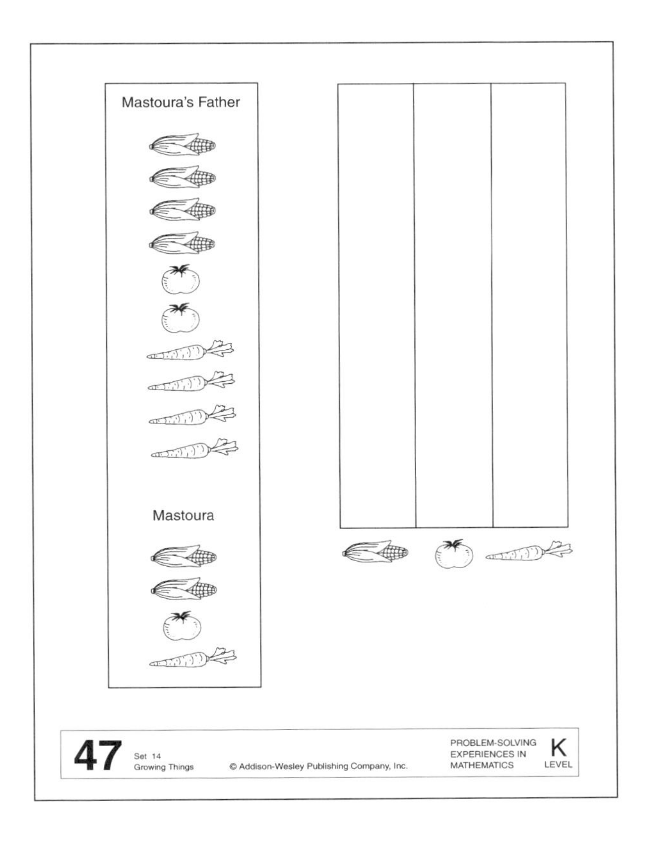

MATERIALS

interlocking cubes (at least 6 yellow, 3 red, and 5 orange per child or pair)

Understanding the Problem

- What did Mastoura and her father plant? (*corn, tomatoes, and carrots*)
- Did they each plant the same number of rows? (*no*)
- What are we trying to find? (*what kind of vegetable is planted in the most rows, the fewest rows*)

Solving the Problem

- Look at the graph. How are the columns labeled? (*with pictures of corn, tomatoes, and carrots*)
- How could we find out how many rows of corn there were? (*move the yellow cubes from the list to the graph*) Tomatoes? (*move the red cubes to the graph*) Can you fill in the rest of the graph? Make sure children align their cubes at the bottom of the graph.
- How can you tell what kind of vegetable is planted in the most rows? (*It's the column with the longest line of cubes, corn*)
- What else do we need to find? (*what kind of vegetable is planted in the fewest rows*) How can you tell? (it's the shortest column, tomatoes)

STRATEGY ASSESSMENT IDEAS

Listen and watch as children work to see if they

- place the cubes correctly in the graph
- interpret the graph by comparing line lengths to arrive at the correct answers

Solution

Complete a Table or Graph

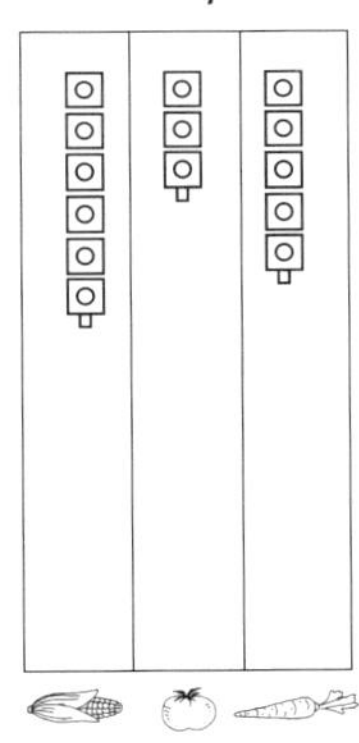

Corn is planted in the most rows. Tomatoes are planted in the fewest rows.

Related Problems: 55, 32, 31

▶ *turn the page*

Problem Extensions

1. Suppose Mastoura planted another row of carrots. What vegetable would they have planted the most of? (*corn and carrots would be equal, 6 rows each*)
2. Have the class make predictions, then add cubes to a class table for their favorite vegetable. What can they tell from the table? Were their predictions right?

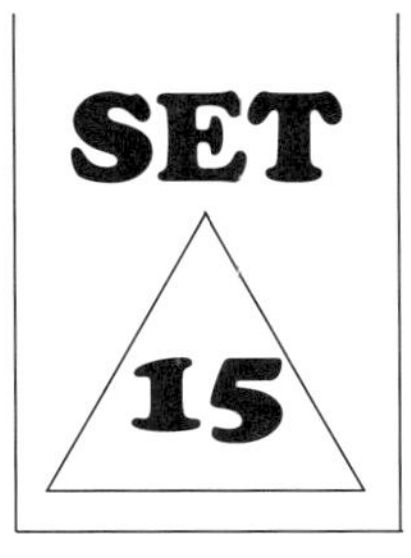

Going Shopping

Megan and Peter were shopping for birthday presents with their big sister, Crissy. Megan was looking for a present for her friend Shondra, who loved purple. Crissy found some purple hair bows, but Megan said that they wouldn't work—Shondra's hair was too short. Megan looked at colored pencils and a soccer ball. Soccer was Shondra's favorite sport, but she already had a ball. She already had colored pencils, too. Nothing seemed right.

Peter was looking for a present for his friend Joe, who liked to read and draw pictures. Peter found some paints, but Joe already had paints. He saw some games and a purple, soccer-ball pencil holder. His friend Joe liked baseball better than he liked soccer, so the pencil holder wasn't right.

All of a sudden Megan and Peter said at the same time, "Here's something for you!" Megan said the colored pencils would be just right for Joe. Peter said the purple, soccer-ball pencil holder would be perfect for Shondra. They were very glad they had come shopping together rather than separately!

Discussion Questions

1. What were Megan and Peter shopping for? (*birthday presents for their friends*)
2. What were some of the things that wouldn't be right for Shondra? (*hair bows, colored pencils, a soccer ball*)
3. What did Joe like to do? (*read and draw pictures*)
4. Why did Megan and Peter say, "Here's something for you!" together? (*they each had found a present for the other's friend*)
5. Have you ever shopped or made a birthday present? Did you ever have trouble deciding on just the right gift?

57 CONCEPT/SKILL ACTIVITY

Given a Picture, Choose an Addition Question

MATERIALS

1" counters, 1" tiles, or 1" construction-paper squares (at least 4 per child or pair)

Picture A

3 wrapped presents on the table, 1 in Shondra's hand—see Blackline Master 49.

Questions for Picture A

1. How many presents are on the table?
*2. How many presents does Shondra have in all?

Picture B

2 wrapped presents on the table, 2 unwrapped presents on the floor—see Blackline Master 49.

Possible Question for Picture B

*1. How many presents did Shondra get?
2. How many presents are left to unwrap?

TEACHING ACTIONS

1. Show and discuss Picture A. Have the children use counters to represent the presents.
2. Read the questions for Picture A. Have the children use the counters to help them choose the addition question.
3. Repeat for Picture B.
4. (*optional*) Have children tell the answer to each addition question.

58 CONCEPT/SKILL ACTIVITY

Tell the Operation

MATERIALS

pennies, 1" counters, or 1" construction-paper squares (at least 12 per child or pair)

Problem A

Megan had 10¢. Then she found 2¢. How much money does she have now?

Problem B

Megan shopped for something to buy with her 12¢. She found a new dinosaur sticker and bought it for 10¢. How much money does she have left?

TEACHING ACTIONS

1. Read and discuss Problem A.
2. Have the children use pennies to represent the action.
3. Ask children which operation is needed to solve Problem A—*putting together* or *taking apart*. Discuss how the action in the story tells which operation is needed.
4. Repeat for Problem B.
5. (*optional*) Have the children solve each problem.

59 PROCESS PROBLEM

After shopping, Crissy took Megan and Peter to the snack bar. There were 2 kinds of fruit drinks, lemon and orange. (Have children color the drinks on the sign yellow and orange.) There were 2 kinds of food: pizza or muffins. (Have children color the pizza red and the muffin brown.) Megan couldn't make up her mind. She found 4 different ways to put a drink with a kind of food. Can you find the 4 ways?

MATERIALS

yellow, orange, red, and brown 1" counters or 1" construction-paper squares (at least 3 of each per child or pair); yellow, orange, red, and brown crayons

Understanding the Problem

- How many flavors of drinks were there? (*2; lemon and other orange*)
- How many kinds of food were there? (*2; pizza and muffins*)
- How many different ways did Allison find to put a drink with a kind of food? (*4*)
- What are we trying to find? (*the 4 different ways*)

Solving the Problem

- Suppose you started with a lemon drink. Put a yellow counter on one of the glasses to show a lemon drink. What kind of food can you choose? (*pizza or muffin*) Use a red counter for the pizza, or a brown counter for the muffin to show the food you chose to go with the lemon drink.
- Suppose you chose a lemon drink again. Put a yellow counter on another glass. What food can you choose? (*whatever wasn't chosen the first time*) Show that with a red or brown counter.
- If we choose lemon again, is there a different food we could use? (*no*) Can you find the last 2 pairs?
- When you're sure you have 4 *different* ways, color the glasses and draw the pizzas and muffins to match the counters and links.

STRATEGY ASSESSMENT IDEAS

Listen and watch as children work to see if they

- name drink and food combinations
- organize the search for combinations (e.g., find all combinations with the lemon drink first)
- find all possible combinations

Solution

Complete an Organized List

The combinations are lemon drink and pizza, lemon drink and muffin, orange drink and pizza, and orange drink and muffin.

Related Problems: 36, 35

Problem Extensions

1. Suppose the drinks were lime and grape and the foods were tacos and fruit salad. How many ways could Megan put them together? (*4*)
2. If there were another drink, grape, how many ways can she put the drinks and food together? (*2 more ways, or 6 in all*)

60 PROCESS PROBLEM

Crissy helped Peter find 2 pairs of shorts and 2 new T-shirts for the summer. The shorts were red and blue. One T-shirt had a red design, and the other had a blue design. Crissy said that Peter now had 4 ways to wear his new clothes. Can you find the 4 ways to put a pair of shorts with a T-shirt?

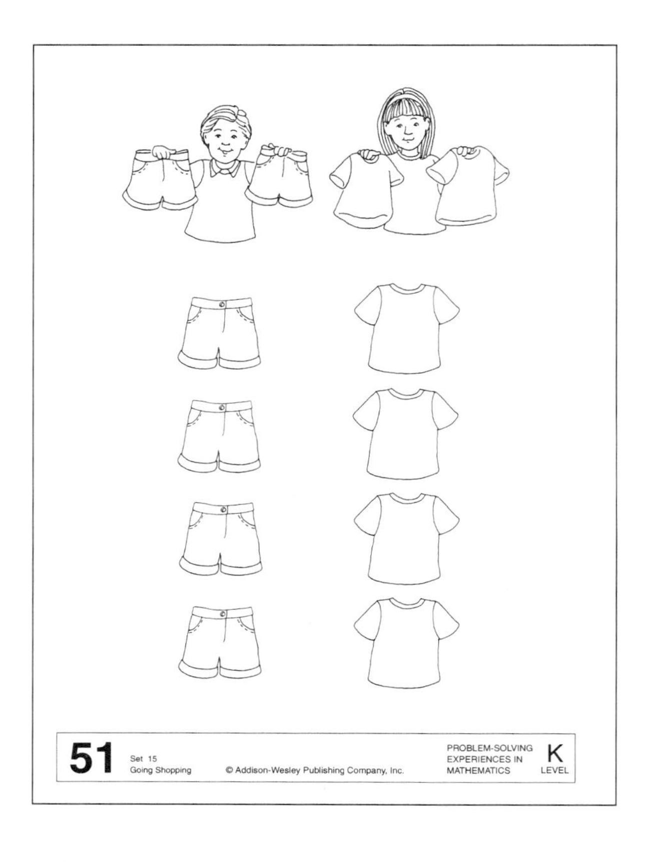

MATERIALS

red and blue 1" counters or 1" construction-paper squares (at least 4 of each per child or pair); red and blue crayons

Understanding the Problem

- How many colors of shorts does Peter have? (2) Color the shorts Peter is holding red and blue.
- How many colors of shirt designs does he have? (2) Color the shirts Chrissy is holding red and blue.
- How many different ways did Crissy find to put a pair of shorts with a T-shirt? (4) What are we trying to find? (*the 4 different ways*)

Solving the Problem

- Suppose you started with red shorts. Put a red counter on one of the pairs of shorts. What color T-shirt can you choose? (*red or blue*) Put a red or blue counter on the T-shirt to show your choice.
- Suppose you use the red shorts again. Can you use the same T-shirt with it? (*no*) Why not? (*there are 4 different ways; this would be the same as the first way*) Put counters to show the shorts with a different T-shirt.
- If we use the red shorts again, is there different T-shirt we could use? (*no*) Can you find the last 2 pairs?
- When you're sure you have 4 different ways, color the shorts and T-shirts.

Solution

Complete an Organized List

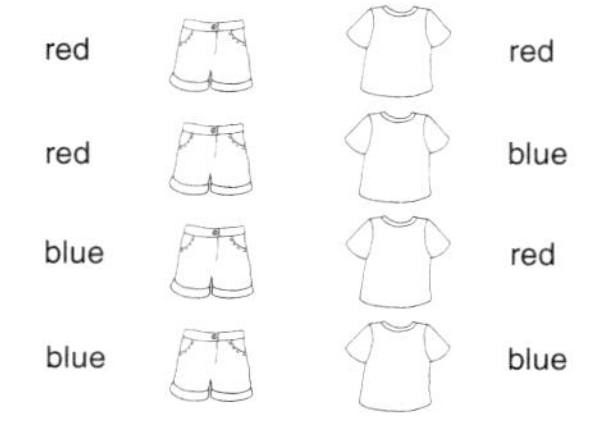

The combinations are red shorts and red T-shirt, red shorts and blue T-shirt, blue shorts and red T-shirt, and blue shorts and blue T-shirt.

Related Problems: 59, 36, 35

Problem Extensions

1. Suppose the T-shirts were white and green. How many ways could Peter put the shorts and T-shirts together? (4)
2. Suppose Peter buys another T-shirt, a yellow one. How many ways can he put the shorts and T-shirts together now? (*2 more ways; 6 in all*)

STRATEGY ASSESSMENT IDEAS

Listen and watch as children work to see if they

- name shorts and T-shirt combinations
- organize the search for combinations (e.g., find all combinations with the red shorts first)
- find all possible combinations

ASSESSMENT APPENDIX

This appendix contains four tools to help you assess your students' progress.

Strategy Implementation Checklist

The Strategy Implementation Checklist contains characteristics of student performance related to each problem-solving strategy. These characteristics can be used to assess student progress over time in their ability to use strategies appropriately. Also, the specific performance characteristics can be used to analyze which aspects of implementing particular strategies students can carry out and which they cannot carry out.

Problem-Solving Observation Checklist

The *Problem-Solving Observation Checklist* includes general problem-solving behaviors and dispositions to be observed and analyzed over time. The first three items address students' selection and use of problem-solving strategies. Item 4 refers to the general approach students use to solve problems, and items 5 and 6 refer to students' dispositions related to solving problems. This checklist can be used as you observe students working in groups solving problems and as you analyze student work on a problem and reflect on their behavior and dispositions.

Focused Holistic Assessment Rubric

The five-level *Focused Holistic Assessment Rubric* is a holistic system for assessing written work, including students' written solutions to problems. This assessment method is a *holistic* method because it focuses on the total solution. It is a *focused* method because one number is assigned to a student's work according to specific criteria related to the thinking processes involved in solving problems. To use the rubric, begin by asking whether the student's paper meets any of the criteria listed under *4 Points.* If so, assign that paper 4 points. If not, move on to the *3 Points* category, and so on.

Mathematics Portfolio Profile Checklist

The intended purpose of a portfolio determines its contents and the method used to assess it. One common use of portfolios is as a collection of student work that can be analyzed for growth over time—that is, to give you a profile of the student's growth in mathematics. It is best to use a small number of criteria to develop such a profile. The following three criteria for the evaluation of portfolios are particularly appropriate for the primary grades:

- the ability to engage in problem solving and mathematical reasoning
- the use of oral, written, and visual modes—approaches or methods of doing mathematical activities—to describe mathematical concepts and relations
- the development of healthy dispositions toward mathematics

The *Mathematics Portfolio Profile Checklist* is designed with these three criteria in mind.

Strategy Implementation Checklist

Strategy	Skill															
MAKE A TABLE/ LOOK FOR A PATTERN	PLACES ITEMS CORRECTLY IN THE TABLE															
	USES A PATTERN TO CORRECTLY EXTEND THE TABLE															
	INTERPRETS THE TABLE TO ARRIVE AT THE CORRECT ANSWER															
LOOK FOR A PATTERN	DESCRIBES THE PATTERN FORMED BY INFORMATION IN THE PROBLEM															
	EXTENDS THE PATTERN CORRECTLY															
	USES THE PATTERN TO ARRIVE AT THE ANSWER															
DRAW A PICTURE	DRAWS APPROPRIATE PICTURES TO REPRESENT INFORMATION IN THE PROBLEM															
	USES PICTURES APPROPRIATELY															
	GIVES APPROPRIATE REASONS FOR USING PICTURES															
GUESS AND CHECK	MAKES GUESSES THAT INDICATE AN UNDERSTANDING OF THE PROBLEM															
	USES PREVIOUS GUESSES TO MAKE BETTER GUESSES															
	CHECKS GUESSES USING THE INFORMATION GIVEN IN THE PROBLEM															
	GIVES APPROPRIATE REASONS FOR GUESSES															
MAKE AN ORGANIZED LIST	CREATES CORRECT ENTRIES FOR A LIST															
	ORGANIZES ENTRIES IN THE LIST															
	LISTS ALL POSSIBLE ENTRIES															
USE LOGICAL REASONING	USES A PLAN FOR RECORDING REASONING															
	CORRECTLY USES ALL CONDITIONS GIVEN IN THE PROBLEM															
	ARRIVES AT CORRECT CONCLUSIONS THROUGH REASONING															
	STUDENT															

Problem-Solving Observation Checklist

STUDENT ______________________________

DATE ________________

	Frequently	Sometimes	Never
1. Selects appropriate solution strategies.	________	________	________
2. Accurately implements solution strategies.	________	________	________
3. Tries a different solution strategy when stuck (without help from the teacher).	________	________	________
4. Approaches problems in a systematic manner (clarifies the question, identifies needed data, selects and implements a solution strategy, checks solution).	________	________	________
5. Shows a willingness to try problems.	________	________	________
6. Demonstrates self-confidence.	________	________	________

STUDENT ______________________________

DATE ________________

	Frequently	Sometimes	Never
1. Selects appropriate solution strategies.	________	________	________
2. Accurately implements solution strategies.	________	________	________
3. Tries a different solution strategy when stuck (without help from the teacher).	________	________	________
4. Approaches problems in a systematic manner (clarifies the question, identifies needed data, selects and implements a solution strategy, checks solution).	________	________	________
5. Shows a willingness to try problems.	________	________	________
6. Demonstrates self-confidence.	________	________	________

Focused Holistic Assessment Rubric

4 POINTS
These papers have any of the following characteristics:

- The student made an error in carrying out an appropriate solution strategy. However, the error does not reflect misunderstanding of the problem or lack of knowledge of how to implement the strategy, but is a copying or a computational error.
- The student selected and implemented appropriate strategies and gave the correct answer in terms of the data in the problem.

3 POINTS
These papers have any of the following characteristics:

- The student implemented a solution strategy that could have led to the correct solution but misunderstood part of the problem or ignored a condition in the problem.
- The student applied an appropriate solution strategy, but
  - **a.** answered the problem incorrectly for no apparent reason,
  - **b.** gave the correct numerical part of the answer but did not label it or labeled it incorrectly, or
  - **c.** gave no answer
- The student gave the correct answer and apparently selected appropriate solution strategies, but the student's implementation of the strategies is not completely clear.

2 POINTS
These papers have any of the following characteristics:

- The student used an inappropriate strategy and obtained an incorrect answer but showed some understanding of the problem.
- The student applied an appropriate solution strategy, but
  - **a.** did not carry it out far enough to find the solution (e.g., the student made the first 2 entries in an organized list), or
  - **b.** implemented the strategy incorrectly, leading to no answer or to an incorrect answer
- The student successfully reached a subgoal, but could go no farther.
- The student gave the correct answer, but
  - **a.** the work is not understandable, or
  - **b.** no work is shown

1 POINT
These papers have any of the following characteristics:

- The student made a start toward finding the solution—beyond just copying data from the problem—that reflects some understanding of the problem, but the approach used would not have led to a correct solution.
- The student began with an inappropriate strategy and did not carry it out, with no evidence that the student turned to another strategy.
- The student apparently tried to reach a subgoal but never did.

0 POINTS
These papers have any of the following characteristics:

- The student left the paper blank.
- The student simply recopied the data in the problem, but either did nothing with the data or did something that appears to show no understanding of the problem.
- The student gave an incorrect answer and showed no other work.

Mathematics Portfolio Profile Checklist

STUDENT ______________________	Substantial Growth	Some Growth	No Growth
PROBLEM SOLVING AND REASONING			
1. Understands information given in problems	______	______	______
2. Applies strategies to solve problems	______	______	______
3. Draws logical conclusions and gives reasons for them	______	______	______
ORAL, WRITTEN, AND VISUAL MODES			
1. Relates manipulatives, pictures, and diagrams to mathematical ideas and situations	______	______	______
2. Relates everyday languages to mathematical language and symbols	______	______	______
MATHEMATICAL DISPOSITION			
1. Has confidence in ability to do mathematics	______	______	______
2. Takes risks exploring mathematical ideas and trying alternatives	______	______	______
3. Perseveres in mathematics activities	______	______	______
4. Is interested in doing mathematics	______	______	______

STUDENT ______________________	Substantial Growth	Some Growth	No Growth
PROBLEM SOLVING AND REASONING			
1. Understands information given in problems	______	______	______
2. Applies strategies to solve problems	______	______	______
3. Draws logical conclusions and gives reasons for them	______	______	______
ORAL, WRITTEN, AND VISUAL MODES			
1. Relates manipulatives, pictures, and diagrams to mathematical ideas and situations	______	______	______
2. Relates everyday languages to mathematical language and symbols	______	______	______
MATHEMATICAL DISPOSITION			
1. Has confidence in ability to do mathematics	______	______	______
2. Takes risks exploring mathematical ideas and trying alternatives	______	______	______
3. Perseveres in mathematics activities	______	______	______
4. Is interested in doing mathematics	______	______	______